Regional Development and the Local Community:
Planning, Politics and Social Context

Regional Development and the Local Community: Planning, Politics and Social Context

CLYDE WEAVER
University of British Columbia

JOHN WILEY & SONS

Chichester · New York · Brisbane · Toronto · Singapore

Library of Congress Cataloging in Publication Data:
Weaver, Clyde
  Regional development and the local community: planning,
politics, and social context.
  Bibliography: p. 164
  Includes indexes.
  1. Regional planning—History.   2. Regionalism—History.
I. Title.
HT391.W355   1984   361.6′09   83–21640

ISBN 0 471 90067 2

British Library Cataloguing in Publication Data:
Weaver, Clyde
  Regional development and the local community:
  planning politics and social context.
  1. Regional planning
  I. Title
  361.6   HT391

ISBN 0 471 90067 2

Phototypeset by Inforum, Ltd., Portsmouth
Printed by the Pitman Press, Ltd., Bath, Avon.

To John with thanks

———————

*For Harvey S. Perloff
and François Perroux*

Contents

Preface

This project began embarrassingly long ago: sometime in 1976. It started in Los Angeles and then moved on to London and Paris and Aix-en-Provence—finally to return to LA and to be finished in Vancouver. Much of my early inspiration and support came from John Friedmann who suggested the undertaking and tried to force me to keep it within bounds. Celia Carden, then at John Wiley, Chichester, trusted me to create a digestible essay. For all this I am grateful. The end result seems meagre after so much time and help.

The book is a social and economic history of ideas about regional development and planning, written in the form of an argument. The outline is presented in Chapter 1, but I should provide a few reflections on what I feel are its strengths and shortcomings. My inclinations have always run towards history and philosophy. Some may feel that what follows is indeed historicism spiced with a sprinkling of metaphysics. (I can't really apologize. Although pedigreed as a geographer and planner the other tendencies refuse suppression, thank goodness.) The strongest points about the book are probably, first, the historical synthesis presented, especially the analysis of libertarian contributions to planning origins and the notion of regional development. And second, though more tenuous no doubt, is the related theorizing in the last chapter which suggests a rationale for grass-roots reconstruction of the social economy in Western industrialized countries like France, Britain and the United States. Ordinary people have not waited for intellectual approval to start finding ways to cope with contemporary restructuring of the global economy. But a relevant contextuating metaphor is essential to help build momentum and gain cultural recognition.

Timing may be the book's general weakness. The Zeitgeist has been transformed since the turn of the 1980s. I would argue that the apparent trajectory of events remains the same, but I may be writing here in an earlier idiom. If I were starting again now, because of my own changing attitudes, I would probably aim for a less single-minded interpretation. I've come to believe that we shouldn't flatten the world with consistency. A multidimensional representation is necessary—an eclectic balancing and bounding of relevant information. Social theories such as models of socioeconomic development are myths which provide a 'text' that might be actualized in appropriate situations. Ascertaining a reasonable approximation for a given setting requires a reorientation of our methods for the creation and validation of social knowledge.

The construction I end with here in Chapter 8 has a simpler substantive problem. My purpose throughout the essay was to establish direct historical links between regional development and local community action, and to argue for the appropriateness of grass-roots approaches to recreating local well-being in the present international context. The internal logic of such a proposition, presented as it is, made it difficult for me to go on to elaborate the inter-connections and interface between bottom-up development initiatives and the macroscale managerial strategies of government and the corporate sector. The linear opposition set up strikes me as unrealistic in retrospect. Obviously both occur simultaneously in existential terms. And while their dynamics can appear as mutual negations, in the phenomenological world they coexist in a complex web of contradictory and complementary relations. What seems necessary to take advantage of both is a yin–yang theory of societal development. That is something for the future however. Currently intellectual legitimation in the social sciences still rests upon the century-old German dispensation of the second industrial revolution and the simplistic post-War Anglo-American tradition of neo-positivist empiricism.

I owe Christine, Kenneth and Heather-Anne uncounted apologies for my preoccupations of the last decade. Joanne Jessop, my colleague and frequent collaborator over the past two years, helped me regain the intellectual interest to get on with my work.

CLYDE WEAVER
Monterey Peninsula, California
February 1984

CHAPTER 1

Rethinking the Regional Question

Since the widespread unrest and violence of the late 1960s there has been a growing ferment within the social sciences. This has been particularly true of applied fields such as regional planning, where self-examination has increasingly taken the form of far-ranging criticisms of received theories and planning strategies. While theorists have argued among themselves, however, political events have passed them by. Governments in many Western countries have all but abandoned their former commitments to regional development policy, and in several areas militant regionalist movements have appeared, challenging the very right of central governments to preside over regional affairs.

Clearly there seems to be a need for rethinking the regional question and all it entails, from the theories of regional science to the role of politics in development. This study is a contribution to that effort. What follows is an interpretive essay on the social and intellectual evolution of regional planning in France, Britain, and the United States, beginning with the socioeconomic conditions and theorizing of the nineteenth century and ending with some speculative considerations on appropriate planning strategies for the 1980s. My central argument is that early regional planning, during the 1920s and '30s, proved unsuccessful because of the economic and political relations of the period, as well as the fact that planners lost sight of their central objective: local political control of economic power. With the rebirth of regional planning in the 1960s the very possibility of independent action was denied. Planning doctrine became an extension of the ideology of multinational capitalism. People and places that did not fit the scheme of the corporate economy found little guidance in its pronouncements. Today, though, because of a growing divergence between the interests of corporate economic power, local communities, and central governments, grassroots regional planning may represent a viable alternative to arbitrary corporate-bureaucratic control. I will argue that contemporarily in industrialized countries regional planning should be one of the central strategies of democratic action.

This analysis rests on examination of the systematic, cumulative changes in three overlapping sets of historical and geographic relationships: (1) the transformation of contradictions between cities and the surrounding countryside,

1

(2) the changing form of class conflict and class alliances in industrial society, and (3) the evolving links between industrial capitalism and political sovereignty. At a more general level, I have found it useful to conceptualize these changes as transformations in the political and economic relations of capitalist urban-industrialization.

1. REGIONAL PLANNING AND THE CONTRADICTIONS BETWEEN CITY AND COUNTRYSIDE

One of the most striking characteristics of industrial society during the last two centuries has been the profound transformation of rural/urban relations. The relatively balanced economic and political ties between cities and their surrounding regions in pre-industrial Europe and North America were progressively undermined by the development of urban-industrialization (see Chapter 2). After a brief incubation period in rural England, industrial technology and capitalist modes of industrial production moved to the city. This was a decisive step in freeing both capital and labour from the laws and cultural norms of traditional territorial communities. Its predominant social form was the *factory system* of production. By the mid-1800s people and machines were rapidly being crowded into large urban workshops, where hours of work could be closely regulated and new, more efficient manufacturing techniques could be introduced.

For the first time, perhaps, productive activity was definitively removed from the home environment; most vestiges of family economic organization and domestic production relations were suppressed for many decades. Both the small-scale territorial community and the biological family were alienated from the production process, and people in ever-increasing numbers were pushed off the land to become urban factory workers. This freed the countryside as well for organization along factory lines. The metropolis boomed, becoming itself an extended factory—surrounded by the squalor of working-class housing. The countryside was first reduced to a position of political and economic subservience, and then, in much of the industrial West, it was all but obliterated as a social environment. Farms increasingly came to resemble large, open-air factories, owned by absentee proprietors and manned at the appropriate seasons by a few transient labourers. This process was entering a critical stage in many areas by the end of the First World War.

Regional planning, in its classic form, was first and foremost a response to the metropolitan explosion. Planners like Lewis Mumford and Howard Odum wanted to stop the flood of 'metropolitanization' and begin a reconstruction of regional life. Drawing on the ideas of earlier regional activists, civic reformers, and libertarian socialists, they advocated a revitalized territorial civilization built around largely self-sufficient regional communities. Decentralized industry and a new balance between town and countryside was to be made possible by the use of appropriate technology, such as rural electrification, applied higher education, and widespread highway construction (Friedmann and Weaver, 1979, Chap. 2).

Regional planners of this first generation felt that rural/urban relations could be transformed by the rational application of science to society (see Chapter 4). Their analyses suffered from several fatal weaknesses, however. First, they seriously underestimated the role of structural economic relationships in creating metropolitan dominance. In the heyday of ascendent capitalist industrialism, they failed to recognize—or recoiled from acknowledging—that the destruction of territorial communities was an integral outcome of capitalist economic relations. In an integrated economic system that encourages free movement of all the factors of production and defines all values in terms of commodity relations, territorial communities found themselves locked in a head-on competitive struggle. Places enjoying strategic advantage, for whatever historical reasons, had a commanding lead.

This inherent spatial conflict in market economies, as well as the corollary struggle between different social classes (Marx and Engels, 1978), was a second major oversight of planning theorists. As liberal social reformers, they down-played the central contradictions of industrial capitalism and refused to elaborate on their implications for regional reconstruction. Like the French regionalists of the late nineteenth century, they argued that regional devolution posed no political threat to the capitalist economy or the centralized nation-state. On the contrary, they called on the central government to help them in their efforts, arguing that 'regional balance' was in the best interest of the country as a whole.

Such theories not only missed the necessary link between capitalist industrialization and geographically uneven development but they completely distorted the relationship between government and the dominant economic classes. Following the lead of English Fabians and American Progressives, regional planners argued that incremental reforms through enlightened government could bring an end to the worst evils of urban-industrialization. Metropolitan growth could be reversed by institutional intervention in the public interest. When planners took a hand in governmental policy formulation during the Great Depression, however, these hopes were sadly disappointed. Rather than stopping metropolitan expansion, programmes like the Tennessee Valley Authority and the Greenbelt New Towns proved a potent force for encouraging industrial growth and fueling urbanization (Friedmann, 1955; Friedmann and Weaver, 1979, Chap. 3).

The Second World War brought an end to serious experimentation on the domestic front. In England the Barlow Report was shelved until after the War and in the United States a nervous Congress voted to dismantle the National Resources Planning Board in 1943. When regional planning reemerged on the international scene in 1960, it was an entirely different vision which provided its theoretical basis (see Chapter 5). Economic development was the *leitmotif* of the 'American Century', and regional planning was a method for bringing economic expansion to places which had been by-passed by unguided market forces. The contradictions between town and countryside, which had provided the stimulus for earlier regional planners, were relegated to the dustbin of professional thinking. In fact, regional development became a matter of spreading the gospel of urban-industrialization.

A new generation of theorists arose, more in tune with the ideological assumptions of their era. For them, regional planning became an exercise in *spatial development*. The planner's task was not to halt metropolitan development, but rather to encourage it. Writers like François Perroux, Douglass North, John Friedmann, and Walter Isard proposed a strategy of *polarized development*, based on extending the sphere of metropolitan dominance. Corporate industry would create a more even geography of employment, income levels, and living standards through complete integration of the space economy. Leading industries would spark the process of regional growth, causing expansion in the most dynamic urban centres. Economic stimuli would spread throughout the urban system as corporate production linkages penetrated the national periphery. The job of regional planning was to provide the necessary infrastructure and public guidance to speed this process along. Depressed areas would share in the general prosperity through establishment of designated *growth centres*, where corporate investors could be aided by government to create industrial plants which would receive the *trickle-down* effects of national growth. Understanding the logic of industrial location became the key to regional planning (Friedmann, 1964; Friedmann and Alonso, 1964).

There can be little doubt that regional economic policies based on such theories helped in some measure to expand the geographic sway of corporate industry during the 1960s (see Chapter 6). This time planners were working *with* the dominant trends in economic production rather than against them. An examination of statistical data for the United States and western Europe during this period shows that the growing metropolitan centres were absorbing a majority of the working-age population and that the corporate sector accounted for a staggering proportion of jobs and economic production (Berry 1973a; Hall and Day, 1978).

There was, however, another side to the picture. Metropolitan growth brought in its train an unexpectedly severe set of social problems: physical expansion and decay, fluctuating job opportunities, race riots, and, eventually, fiscal crisis (Banfield, 1968; Castells, 1972; Harvey, 1973; O'Connor, 1973). Planned growth centres in non-metropolitan areas did not seem to grow, at least not as planned (Pred, 1976). The kinds of industries which public location controls and investment incentives were able to attract to peripheral locations did little to generate further growth. They tended to be low-skill assembly operations which had feeble income and employment multipliers and few linkages to the surrounding regional economy. The dynamic stages of production—managerial decision making and R & D—stayed in the metropolis (Massey, 1978b). Agriculture was turned into a corporate enterprise as well, as the countryside emptied and the family farm was replaced by agribusiness.

Even farm workers were being displaced by rapid mechanization (Buttel and Newby, 1980).

Spatial planning proved effective in helping to *rationalize* the geographic structure of industry, but gave little assistance to territorial communities that did not fit the design. Polarized development continued to create strikingly uneven patterns of social and geographic income distribution. What is more, as people moved away from the smaller towns and farming areas, formerly viable regional economies became dependent on outside decision makers. Dismantlement of the rural economy, based on the collapse of agricultural service activities, competition from large national and multinational corporations, and independent corporate location decisions took on the air of a process of underdevelopment. This was the reverse of the normative predictions of regional planning theory, as a new and more specialized spatial division of labour made economic *interdependence* seem more like economic *dependence* (see Chapter 7). Not even the large cities proved immune; jobs could be moved by globally extended multinational corporations from one region to another, or from one country to another, in search of political security, cheap labour, and a 'good business environment' (Holland, 1976a; Mandel, 1975).

By the mid-1970s 'stagflation' (i.e. recession plus inflation) had become generalized throughout the world economy, partially attributable to the investment and pricing practices of the multinationals. Led by the OPEC cartel and big oil, oligopolistic market control caused real living standards to drop in even the most favoured metropolitan areas. Consumer goods, housing costs, and basic service expenditures skyrocketted, followed by tax revolts, cutbacks in public services, and temporary rent freezes. Recent population trends in the United States, for instance, show that people have even started to move back to non-metropolitan areas in significant numbers (Fuguitt, Voss, and Doherty, 1979). The dependent non-metropolitan economy, though, offers only limited scope for employment expansion along traditional lines. In addition, regional planning of the 'spatial development' school holds small hope for solving such dilemmas, as corporate economic power seems to have passed beyond the bounds of territorial political control. Today, the most powerful national governments find it increasingly difficult to control the actions of multinational corporations, even when they are inclined to do so. The historical contradictions between town and countryside have taken on a markedly different form; first the countryside was made all but uninhabitable as a human social environment, and then the economic and political viability of all types of territorial communities—including the metropolis—came progressively to be threatened by multinational corporate 'independence'. The metropolis no longer dominates the countryside in the traditional sense. They are *both* held in fief by the footloose factory system of production, embodied in the multinational corporation (Hymer, 1968, 1972a).

These transformations have called into question the most fundamental concepts of regional theory as inherited from the 1950s and '60s, and caused an increasingly heated debate among planners (see Chapters 6 and 7). A first approximation of their underlying dynamics, however, can be seen in relation to changes in two other dimensions of industrial society: class relations and political sovereignty.

2. URBAN CHANGE, CLASS STRUGGLE, AND CLASS ALLIANCES

Like the geography of rural/urban dominance, the class structure of advanced industrial society has developed out of production relations initiated at the industrial revolution. Transformations in one have closely paralleled and interacted with transformations in the other, both reflecting basic changes in the capitalist economy (Soja, 1980). Historically, regional theorists have had little interest in these vital linkages.

In Britain an industrial working class was created by pushing peasant farmers off the land, through such devices as the Enclosure Acts and the Poor Law (Hobsbawn, 1962). Labourers—landless, hungry, and desperate—were needed to make the factory system work. Rural piece-work occupations like handloom weaving, part of the transitional *putting-out system*, were suppressed by plummetting piece rates and, finally, the King's soldiers. In the United States, from the Revolution until the Civil War, these same European immigrants provided the needed mill hands (Beard, 1924; Carroll and Noble, 1977). After the 1860s they were supplemented by American farmers pushed off the land by mechanization (Conrat and Conrat, 1977). In France a tenacious peasantry hung on to its small parcels of land, but in Paris and textile centres such as Lille and Lyons a genuine industrial working class was created (Braudel, 1973; Hobsbawm, 1962).

Factories, working people, and industrial cities were all brought into being during the eighteenth and nineteenth centuries by the same set of processes: the rise of capitalist industrialism. It was the *conditions of everyday life* of this new proletariat in the new industrial cities which appalled and outraged a small minority of the educated classes. (Working conditions were equally dehumanizing, but they were less visible.) Progressive reform, urban religious movements, and radical socialism sprang up at the interface of working-class misery and middle-class uneasiness. Urban and regional planning in the contemporary sense was a liberal-leaning mixture of all three elements, becoming firmly established among the new 'urban' professions over the half-century between 1880 and 1930 (Heskin, 1980). Their earliest motives were concerned with reforming living conditions of the industrial working class and controlling the growth of industrial cities.

The precursors of regional planning—with much clearer insight than most of their successors—saw the connection between economic relations and urban living conditions. They advocated starting over again on a small scale in new locations. The nineteenth century utopias of Robert Owen (1972), Charles Fourier (1971), and Ebenezer Howard (1946) all envisioned new, freestanding industrial communities which, in varying degrees, would restructure the economic and social relations of the industrial city. Experimentation followed on both sides of the Atlantic. As adopted by the Regional Planning Association of America during the 1920s, however, the 'new town' idea had already lost most of its social and economic content, becoming fundamentally a physical planning concept (Mumford, 1925a) (see Chapter 3). By the late 1940s, the famous transatlantic 'garden city' debate between Lewis Mumford and Frederick Osborn had degenerated into an argument over the style of housing units and urban population densities (Hughes, 1971).

Such theories, experiments, and debates had little impact on the evolution of industrial capitalism. In the first quarter-century after the Second World War the processes of urbanization and factory organization of the economy were nearly completed in western Europe and Anglo-America. Until almost the end of this period, not only did the contradictions between town and countryside fade from view, but living conditions of the urban working class came to be seen as a direct outcome of economic growth. The idea of a correlation between corporate expansion, metropolitan growth, and urban living conditions was widely accepted. In large measure, urban-based labour unions as well as business interests, large and small, subscribed to this notion. Labour/ management alliances were struck (Holland, 1978) and, as I have already argued, regional policy became one of the accepted governmental approaches to aid in their realization.

In the late 1960s this apparent harmony of interests was burst by urban riots in the United States, near governmental collapse in France, and the growing 'intransigence' of British labour unions. Although a whole generation had grown up during the 1950s and '60s believing in the hope of increasing prosperity and universal betterment, the realities of urban life were becoming unavoidable. Corporate expansion was achieved in large part through sacrifices by the majority of the population, and new, rationalized patterns of capital investment were actually taking jobs away from members of the cooperating labour unions. Metropolitan expansion and related public investment did much to accelerate the circulation of capital and increase corporate profits, but in both social and geographic terms the costs and benefits from such expansion were very unevenly distributed (Harvey, 1973). The outflow of manufacturing jobs from large cities, subsidized by policy incentives, had a problematic effect on living standards in peripheral areas (Richardson, 1978a, Chap. 10), but their net impact on working people in metropolitan centres was indisputable: increased structural unemployment.

Economic transformation of the metropolis in 'post-industrial' society is a widely recognized phenomenon, discussed by analysts with quite divergent ideological perspectives (e.g. Berry, 1973b; Castells, 1978). Conceived as part of the more general economic and political crisis of the 1970s and '80s, structural changes in the metropolitan economy are now clear in their main outlines (Carney, Hudson, and Lewis, 1980; Crisis Reader Editorial Collective, 1978; Poulantzas, 1976b).

New industrial location patterns, including disinvestment in many traditional metropolitan centres, have created a striking setting for conflict. Organized labour, historically centred around secondary industry, finds itself faced with a direct challenge on the basic issue of merely maintaining employment levels (Goodman, 1979). This is especially true in aging industrial zones such as Nord in France, the English North East, and the 'Manufacturing Belt' in the United States (e.g. Damette and Poncet, 1980; Rifken and Barber, 1978; Sternleib and Hughes, 1975). As secondary industry moves out of the metropolis, the well-documented decay of the central city and older suburbs continues, despite the inroads of 'gentrification'. Affordable working-class housing stock in the metropolis decreases. New job creation takes place almost exclusively in a few advanced-technology manufacturing sectors, high-level tertiary activities, and poorly paid consumer services (Lipietz, 1979a).

The result of these changes has been a major reorientation of the socio-economic and geographic structures of the economy. Increasingly, big cities have become the domain of highly paid, well-educated professional people and those who provide for their immediate service needs. Even as industrial workers are forced to leave in search of alternative employment opportunities, living costs continue to escalate dramatically for those remaining. Standards of living are visibly eroded. Fewer resources flow into the metropolis for trans-formation, now typically done in cheaper, highly specialized rural locations and abroad in the Third World. The city becomes increasingly dependent in material terms on the countryside, but rural production centres wield almost no control over production decisions. Many cities face fiscal crisis (Alcaly and Mermelstein, 1976; Burchell and Listokin, 1980), while non-metropolitan workers have less control over their destinies than their big city predecessors (Malizia, 1978), and resource areas are exploited for their physical assets with hardly a hint of long-term development (Markusen, 1978a). Those who *can*, follow the jobs—towards the Sunbelt in North America (Heil, 1978) and, to a lesser extent, the Mediterranean coast in Europe (Damette and Poncet, 1980; Lipietz, 1979a). 'Tertiarization', however, is producing employment structures in sunbelt areas which closely resemble emerging national patterns (i.e. metro-politan service centres and rural manufacturing). Corporate integration of the economy continues, marked by new and higher degrees of regional specializa-tion and dependence, and a growing proportion of the traditional urban work force displaced as redundant (Carrère, Catin, and Lamandé, 1978; Clark, 1980b).

In this environment several new trends in regional and national class alliances are appearing. The most conventional has both elite and working-class leaders in potential growth areas courting multinational capital invest-ment. This is the old pattern of geographic competition, the only twist here being that 'success' stories are happening most often in novel locations, outside the industrial heartland. A variation of this theme—but a very important one—is the emergence of regional coalitions, often related in some fashion to a regionalist political movement. Such coalitions can vary widely in class com-position—some leaning towards 'regional capitalism', grouping local entre-preneurs and labour against the outside world, and others representing a sort of regional left-populism, searching for an 'alternative path' to development (Markusen, 1978b; Nairn, 1977; Richards and Pratt, 1979). Finally, there are some signs of interregional class alliances between workers in related branches of industry (Hobsbawm, 1977; Markusen, 1978a, 1978b).

Three things are of particular importance in the present context. First, none of these alliances rest on the big capital–central government–national labour agreements characteristic of the 1960s. Second, there are few signs of active organization behind the doctrine that current economic and population growth in sunbelt areas will stimulate positive changes in other regions. And third, while casting further doubt on central governments' ability to deal with regional problems, the new alliances suggest that local approaches to regional development may find substantial grassroots political support. I will develop some of the implications of these trends in Chapter 8.

3. INDUSTRIAL CAPITALISM AND POLITICAL SOVEREIGNTY

At the heart of the regional question is the problem of political sovereignty and its relation to economic production. Modern regionalism as a political tradition in Western countries took root in the eighteenth and nineteenth centuries; it developed concomitantly with consolidation of the nation-state (Gras and Livet, 1977). Along with formation of the urban working class and the growth of industrial cities, regionalism was intimately connected to the establishment of national markets and the expansion of industrial capitalism (see Chapter 2).

Earlier alliances between the rising middle classes and absolute monarchy in Britain and France had created the foundations of the contemporary national state by the turn of the sixteenth century. Over the next two hundred years capitalist interests gradually assumed the mantle of State power. The national territory, extended to a global scale by imperial expansion, provided a free-trade zone where commerce, resources, and, to a lesser extent, labour could move unobstructed by institutional barriers from one locale to another in response to market forces. This sphere of 'free exchange' was, in fact, created and guaranteed by the State. The American constitution of 1789, as interpreted by the Marshall court, assured creation of a similar national economy within the new federal republic by reserving regulation of interstate commerce to the central government.

As long as national economies were organized on a largely pre-industrial basis, however, national markets meant primarily interregional competition between different commercial and finance interests. While traditional trading and banking centres, especially in the national heartland, enjoyed an undeniable advantage and exercised substantial control over the entire system, the geographic division of labour was far less specialized than in later periods and was based on historically defined comparative advantage. Most areas remained self-sufficient in matters of basic commodity production: people ate and drank their own produce and wore their own particular types of clothing. Only selected high-value natural resources and products which developed a wider notoriety could bear the cost of transport in national and international commerce: Camembert and butter from Normandy, tobacco from Virginia, whisky from Scotland. Regional elites in *developed* areas maintained a diminished but continuing parity in terms of economic and political influence.

All this changed after the early stages of industrialization. The Jeffersonian Commonwealth lost its material foundations with the ascendence of industrial capitalism (Beard, 1913; Green, 1977). Economic decisions which affected the allocation of regional resources and the health of regional economies came increasingly to be controlled by nationwide corporate enterprise (Beard, 1924; Hobsbawm, 1962). Economic rents from the use of regional labour and resources were appropriated by industrial entrepreneurs who reinvested their earnings according to a national calculus. As productive facilities were concentrated in a select number of urban locations, these areas grew not only in terms of economic importance but also population. Forced migration brought increasing numbers of workers into the industrial cities, creating the geographic core of the industrial economy and, concurrently, the critical centre of the national market. Peripheral regions became resource hinterlands and sources of surplus labour. National industry found its natural allies at the seat of national political power, the central government.

The exodus of people from rural regions decreased their political influence in the national capital. Fundamental economic decisions were being made in big cities by the 'captains of industry' and finance (Veblen, 1899), and local government had no effective means of coping with unfavourable decisions which came from outside its jurisdictional boundaries. Even within the American federal system, political and economic sovereignty were shifted to the national level by the structural logic of industrial capitalism.

In the last decades of the nineteenth century regionalism had appeared as the inevitable reaction to this sense of powerlessness (see Chapter 3). In France the Midi was a hotbed of regionalist sentiment. Jean Charles-Brun came forward in 1900 to found the French Regionalist Federation, whose Proudhonist manifesto (Charles-Brun, 1911) became an international rallying point (Vigier, 1977). The Celtic fringe of both Britain and France—Brittany, Wales, Ireland, and Scotland—and Pays Basque were alive with memories of past sovereignty (Hechter, 1975). In the United States, populism, a southern agricultural cooperative movement, swept the South and West in protest against outside economic control and the unresponsiveness of the Washington government (Goodwyn, 1978).

The regionalist doctrine was expressed most typically in carefully chosen patriotic language. Regionalism itself was defined within a national context and the sovereignty of the nation-state was too firmly established to be put in question. Regionalists argued that it was in the *national* interest to assure a more even distribution of economic production and political control, and called upon national leaders to reverse disruptive trends towards geographic concentration and polarization (Lagarde, 1977).

Despite, or perhaps because of, the failure of regionalist movements before the First World War their momentum increased during the 1920s (Cash, 1941; Guiral, 1977; Hechter, 1975). In the United States regional planning appeared as a professional and intellectual programme which incorporated much of the regionalist platform (Mumford, 1938; Odum and Moore, 1938) (see Chapter 4). Mixing regionalism with progressive reform strategies, planners attempted to translate essentially political demands into a technical vocabulary. In the process, the fundamental relationship between political and economic self-determination was all but lost. Regional planning during the 1930s, in practice, became a matter of national-level policy which, on the whole, served strictly national purposes (Friedmann and Weaver, 1979) (see Chapter 3).

Three decades later, in the 1960s, spatial development planning had few open ties with political regionalism (see Chapter 5). Localized unemployment was the key issue which reinvoked the need for regional planning, but national corporations and national government were the assumed vehicles of economic decision making. A complicated rationale was devised to explain how continued polarized development would eventually put an end to regional economic problems. As I have already pointed out, however, evaluation of regional policy initiatives by the end of the decade could give only the most problematic evaluation of their success. In the thirty years following the Second World War, direct foreign investment within the organizational structure of corporate enterprise had clearly created an important new economic actor on the international scene, the multinational corporation, and an increasing number of economists and planners began asking in what measure it could be expected to conform to the political dictates of the bounded nation-state (Hymer and Rowthorn, 1970; Michalet, 1976; Perroux, 1969). National sovereignty was being threatened by the continued expansion of functional economic organizations, just as the integrity of smaller territorial communities had been undermined before it. The commonality of interest between national decision makers and corporate executives could no longer be assumed, and the ability of countries to shape their economic future—let alone the destiny of subnational areas—became a matter of increasing concern. By the mid-1970s, policy analysts came to question whether regional economic incentives and controls were even serving national goals (Holland, 1976a) (see Chapter 7).

Today, with evident aggravation of regional economic problems and the apparent decline of national sovereignty, two marked trends can be observed in Western countries. National governments, preoccupied with the international economic situation, have made an aboutface in their earlier support for regional development programmes (Richardson, 1978a). Regional policy has been transformed into an adjunct of short-term, anti-cyclical economics (Passeron, 1979). The bright hopes of the 'great society' era—that regional poverty was a transient phenomenon, soon to be eradicated—have apparently been forsaken (see Chapter 6).

Coinciding with this retreat from regional planning has been the appearance of an increasing number of active regionalist movements. As faith in the ability of national governments to resolve local problems and cope with the tactics of multinational capital has wavered, grievances old and new have served as rallying points for regional discontent. National unity and 'regional balance' are no longer the dominant focus they were at the turn of the century. Demands for devolution and regional autonomy have taken their place. Calls for separatism are even heard. And a permanent secretariat has been established in Strasbourg, under the banner of the *Free European Alliance*, to act as a 'representative of European peoples fighting for their autonomy' (*Le Monde*, 16 July 1981).

In the United Kingdom a home-rule election was held for Scotland and Wales in March 1979. The referendum failed, in both cases, but the ensuing political realignments were severe enough to bring down the minority Labour government, and the economic and fiscal policies of the Conservative Thatcher government have helped to keep the cauldron bubbling. Of a more drastic nature, the bloody question of British rule in Northern Ireland has cost the lives of some 2,000 people over the last ten years, and no resolution is in sight. The nation which rose to world empire on the shoulders of capitalist industrialism may be faced with the possibility of almost complete dismemberment (Nairn, 1977).

In France, also, regionalism is the order of the day (Gras and Livet, 1977). Long-standing separatist movements in Brittany and Corsica continue to mix demands for regional autonomy with escalating violence and terror tactics. On the French/Spanish border Basque separatists continue their strategy of bombing and terrorism. Industrial dislocations in northern steel manufacturing districts have caused violent strikes and armed intervention by the Paris government, while dependent industrialization in the Midi has brought about a rebirth of Occitanian regional sentiment. Even in the touchy border areas of Alsace and Lorraine trends towards regional organization have emerged, built around German-language affinity groups. In this atmosphere, it is not surprising that a strong regionalist plank was incorporated in the electoral campaign of the Socialist government during the 1981 presidential and parliamentary elections (Parti Socialiste Français, 1980). One of Mitterand's first decisions in domestic affairs was to promote regional devolution and special status for Corsica.

Although regionalism in the United States lacks the historical scope and traditional language incentives which characterize western Europe, regionalist movements are also abroad in America. The celebrated sunbelt/snowbelt controversy has taken on political as well as economic dimensions. Workers in the Northeast are attempting to control interregional capital mobility through the leverage of union pension funds (Rifkin and Barber, 1978). Regional blocks of legislators and state governors have formed increasingly vocal and powerful sectionalist groups. In the wake of worldwide energy problems, historical arguments concerning state's rights have gained unexpected momentum—the 'Sage Brush Rebellion'. From New Mexico to Alaska, state governments and local interest groups are preparing to fight the federal establishment over control of public lands and natural resources. Alongside this, community power movements have gained an unprecedented level of political support (Boyte, 1979; Morris and Hess, 1975) and growing Hispanic minorities all across the sunbelt are beginning to exercise a recognizable influence on regional consciousness. In Puerto Rico, one of the testing grounds of US development policy since the 1940s, separatist organizations have strongly contested the idea of statehood (Hume and Shannon, 1979; Maldonado-Denis, 1972; *New York Times*, 22 January 1983).

In both the United States and western Europe national sovereignty is being challenged on two fronts. Multinational corporate power is undermining the nation-state's ability to control economic conditions and dissatisfied regional coalitions are beginning to demand leave in handling their own affairs. While the outcome of these trends is still quite uncertain, their possible ramifications are of extreme significance. The attendant conflicts and upheaval may herald the emergence of a new international political–economic order; they may be only passing signs of structural readjustment. In the context of regional planning, however, they represent a profound challenge to prevailing ideas.

4. OUTLINE OF THE STUDY

In the study that follows I will analyse the historic origins of regional planning and trace its major transformations through the beginning of the 1980s. Then I will go on to suggest a theoretical orientation for planners in Western countries to cope with the changing circumstances of the next decade. Because of their central role in the evolution of regional planning theory and practice attention is focused on three countries: France, Britain, and the United States. While their political institutions and historical experience have varied greatly, their combined contribution to regional planning makes such a comparison essential for an understanding of the contemporary situation.

Chapters 2 to 4 present a sketch of developments from the pre-capitalist period through the Second World War. Chapter 2 analyses the development of industrial capitalism and its impact on urbanization and consolidation of the nation-state. Its central arguments deal with the contradictions between town and countryside, changing relations between the State and different regional elites, and the role of urbanization in the formation of an industrial working class. Chapter 3 outlines the intellectual response to urban-industrialization in the nineteenth and early twentieth centuries. Discussion revolves around the appearance of critical thinking and the social sciences and their relationship to regional planning origins. Chapter 4 gives an account of the first doctrine of regional planning during the 1920s and '30s and analyses its application in planning practice.

Chapters 5 to 7 follow the career of spatial development planning from its intellectual synthesis during the 1950s and early '60s to its adoption as a basis for government policy and its eventual critique and abandonment. Chapter 5 traces the remnants of pre-war ideas as they survived in the British Mark I new town programme, and goes on to describe the evolution of regional science-based theories and their convergence in the concept of *growth centres*. The new town and growth centre approaches are then contrasted to reveal their fundamental incompatibility and the resulting confusion for planners when the two overlapped as a basis for regional policy.

Chapter 6 outlines the application of such policies in France, Britain, and the United States from the early post-war period to the turn of the 1980s. Despite differing circumstances, regional policies adopted in the three countries were very similar, and after two decades of practical experience mainstream evaluations proved inconclusive. By the mid-1970s, however, empirical examination of planning successes and failures seemed somehow tangential, for governments, caught in the throes of global economic crisis, were in headlong retreat from local issues. Regional planning, first legitimated in government circles by the Great Depression (see Chapter 4) was now abandoned, perhaps ironically, for analogous reasons—to be replaced by national-level anti-cyclical policies and post-Keynesian economics.

Beginning in the Third World, however, an entirely different interpretation was put forward to explain this whole chain of events. This *radical* analysis of regional problems has now come to occupy an important place in Western thinking as well. Chapter 7 presents the broad outlines of the radical critique of inherited regional science, from the *underdevelopment/dependency* theories of Latin American writers to Marxist analyses of the *crisis management* strategies of the capitalist State. It is argued that these perspectives provide a more meaningful explanation of regional problems and the fortunes of regional planning than previous formulations, but offer few mid-term alternatives for coping with regions in crisis.

In the last chapter, Chapter 8, I propose a possible strategy for regional development and planning in Western countries during the 1980s. Here the limits of economism are explored and it is posited that both the regional economy itself and participation in the global marketplace offer unrealized opportunities for regional development. The main obstacles are political and ideological, and these can be confronted through a democratic strategy of regional political organization and concerted action.

Thus, I would argue, in an important sense regional theory comes full-circle. As in the beginnings of urban-industrialization, the health and well-being of a local community are seen within the context of the broader political economy, but, to succeed, regional reconstruction must be accepted as fundamentally a local task, based on democratic consciousness, solidarity, cooperation, and struggle.

CHAPTER 2

Cities, Industry, and Empire

When attempting to understand the origins of regional planning it is essential to relate planning and the social theories it has drawn upon to the historical development of industrial society. In this chapter I will present an interpretation of the critical socioeconomic and political changes which occurred in France, Britain, and the United States during the transition to industrial capitalism. It was the interplay of central government power and capitalist modes of factory production which created the new industrial cities, as well as a changing balance between social classes and different geographic regions. These profound changes were at the root of nineteenth century urban and regional social movements, and provided the political and economic foundations for the social theories and criticism which will be discussed in Chapter 3.

The economic relations and dominant institutions of industrial society came to maturity in much of the North Atlantic community in the first decades of the twentieth century. Europe and America arrived at this juncture by different paths, however. In France and Britain it will be necessary to examine the internal transformations of European economic organization and urban life, beginning with the Middle Ages. Urban-industrialization in the United States took place under quite different circumstances. As Douglass North (1955) argued, in North America we find a colonial resource supplier developing to nationhood and industrial prowess outside the framework of traditional European experience. What is remarkable, perhaps, is that urban-industrialization, because of its own structural logic, created many similar circumstances and problems on both continents.

1. THE MEDIEVAL SETTING IN EUROPE

The towns and burgs of the early Middle Ages were not really urban places, although towns often occupied the physical shell of former Roman colonial cities (Pirenne, 1925, Chap. 3; 1958, pp. 197–209). Their primary functions were ecclesiastical and military, their real role in medieval society being protection in times of extreme emergency. Although there were a few larger trading centres, such as Bordeaux, Rouen, and Augsburg, commerce had

15

ceased in any significant sense. Local periodic markets were a mere vestige of earlier institutions, serving the needs of the 'urban' clergy, military, and their servants. Rudimentary urban services and crude craft activities served the same ends. However, there were almost no broader ties between the town and countryside; even political administration had fled to the rural strongholds of the feudality. There was no general urban population in the towns, and the countryside organized economic production as well as political life in nearly self-sufficient demesnes. In the ninth century towns were simple dependent consumers, and many of the foodstuffs which they consumed were even produced within their own ramparts.[1] There has seldom been a more hermetically sealed human environment.

By the thirteenth century Europe had witnessed the rise and fall of four maritime trading 'empires': the Italian city-states (Venice, Genoa, and Pisa), the Lombard League, several generations of Hanseatic trading confederations, and the famous textile cities of Flanders. The close of the fifteenth century saw unification of the first modern nation-states: France in 1480 and England in 1485. But what were the causes of these transformations of feudal society? What was the functional importance of cities? And what were the political and economic bases of the new nation-states?

2. THE TRANSITION FROM FEUDALISM

The communal movement became an important force in Europe during the eleventh and twelfth centuries (Pirenne, 1925, p. 69). There is significant contention as to how this momentous trend began, but Bookchin (1974) identifies two primary causes for the rebirth of urban life: (1) the segmented nature of *feudal political suzerainty*, in which each barony developed its own 'fair' or marginal trading centre, and (2) the competitive settlement of new land, which often required particular concessions and modifications of feudal relations to attract settlers. Feudal obligations such as *corvée* labour were replaced by *in-kind* and even cash payments, and municipal 'freedoms' were guaranteed by legally binding writs and charters (see also Merrington, 1975).

Another related factor, underscored in the famous Dobb/Sweezy debate over the transition from feudalism (discussed below), is the fact that population growth in rural feudal society, after the establishment of relative peace and prosperity, could not be supported by existing agricultural production relations, land availability, and technology. Some of this increase had to be absorbed within an alternative structure if it was to survive. The commune provided such an alternative. As Bookchin (1974, pp. 38–39) notes, the medieval commune was a self-contained centre based on an economic system that satisfied human needs through simple commodity production and local trade. It provided products and skills that could not be acquired in the manorial domestic economy. It was here that independent traders and craftsmen, the precursors of the modern bourgeoisie, were able to flourish. However, as long as the economy was based on simple commodity production and not the accumulation of capital, the medieval commune remained a feudal rather than a bourgeois city.

Among economists speculation about the role of cities in economic history began with Adam Smith (1776); the earlier French 'physiocrats', by virtue of their belief in the unique agricultural origins of wealth, took little or no interest in the question. Smith's well-known argument was that towns acted as the nexus of long-distance trade—as commercial entrepôts—which set the whole wheel of capitalist accumulation and development in motion. Subsequent developments—the qualitative transformation of intraregional trade, the decline of the guild system and rural feudal relations, the ascendance of merchant over landed lords, and, finally, the rise of industry—all this, i.e. the wealth of nations, was a product of the city.

Marx's analysis (Marx, 1857a, 1857b, 1867; Marx and Engels, 1846, 1847) emphasized the importance of internal production relations for the decline of feudalism, *but* he also assigned an extremely important role to merchant capital—the 'first free form of capital'—in the rise of its successor, capitalism. The next influential commentators, Max Weber (1905) and Henri Pirenne (1925), reemphasized the paramount influence of external trade relations on the formation of the bourgeoisie, and centred this transformation around the medieval cities of northern Italy and Flanders—especially Flanders, where the wool trade led to true entrepreneurial competition, industrialism, and urban dominance. Two decades later, Dobb (1946) tried to redirect attention back to internal class relations, but was severely criticized by fellow Marxist Paul Sweezy (1950). The ensuing debate, led off by Sweezy *et al.* (1954), provided little final resolution of the problem, leaving it to Bookchin (1974) and Merrington (1975) to attempt a solution. Both these latter writers emphasized: (1) the crucial *political autonomy of towns,* within the system of parcellized feudal territorial relations; (2) the subsequent transformation of *urban commodity production*, which later became generalized throughout surrounding areas; creating (3) *abstract production, abstract labour*, and *abstract capital* (i.e. the *Market*); as well as (4) *windfall middleman profits*, extracted from functional 'city-regions' and, eventually, *national markets*. The spatial structure of feudal relations allowed existing social structures and processes to be reordered and subordinated to new patterns of organization.

This transformation was achieved through the generalization of commodity relations, which helped destroy the economic basis of seignorial production, and competition among towns to increase their limited monopoly markets. Such increases could only occur, before the rise of entrepreneurial competition, by geographic expansion of sovereignty, which partially explains the urban bourgeoisie's willingness to cooperate with royal pretenders against the fragmented feudality. As well as guaranteeing corporate privileges, through royal protection, creation of a broader national market allowed horizontal expansion of functional economic ties. A national economy was superimposed over the existing regional ones. As for the aspirants to monarchy, alliances with the urban middle classes provided their only possible *domestic* base of economic power, vis-à-vis the claims of the landed aristocracy.

The political ascendancy of the nation-state, as well as the development of true entrepreneurial competition and genuinely national markets, could not be based on 'secondhand profiteering'. The limits of municipal monopoly and the bonds of fragmented feudal sovereignty could only be finally broken by a universal triumph of commodity relations (i.e. the *Market*). This came about through the spread of capitalistic production relations in the *countryside*—the only place where abstract capital and labour were free to compete against the ascriptive political economy of the guilds.

3. INDUSTRIAL CITIES, THE NATION-STATE, AND EMPIRE

Emergence of the nation-state and its establishment as the principal expression of political sovereignty took place over a period of some four hundred years, and involved a confusing quiltwork of alternating alliances and struggle between the urban middle class, monarchy, several segments of capital, the peasantry, and the urban proletariat.

The results of these wars, civil wars and revolutions, which reached a hiatus after 1870, were hegemony of the bourgeois State, formation of the proletariat, the ascendance of finance capital, and complete dominance of the city over the countryside. Capitalist economic relations and State power destroyed the foundations of feudal society and the mercantile city, although their remnants lingered on, periodically reasserting themselves in the form of regionalist movements.[2]

This emerging pattern of national integration and regional specialization was founded on the (vertical) relations between capital and labour and the (horizontal) relations between urban and rural areas. Small-scale, feudal society was superceded by a new mode of relations between people, their work, and their environment, as well as a new set of 'urban' production relations between town and countryside. In Merrington's analysis:

> The dominance of the town is no longer externally imposed: it is now reproduced as part of the accumulation process, transforming and spatially reallocating rural production 'from within'. The territorial division of labour is redefined, enormously accentuating regional inequalities: far from overcoming rural backwardness (seen as a legacy of the past, as in Smith), capitalist urbanization merely reproduces it within specific regions, on a more intensive basis. The creation of the 'reserve army' of cheap labour and the rural exodus could scarcely be seen as 'progress' from the rural standpoint (Merrington, 1975, p. 88).

Mercantile cities had already begun to spur the break-up of feudal society: 'The expansion of the market from a local or regional to an international scale occurred at a tempo that gravely disrupted the harmony of the commune' (Bookchin, 1974, p. 53). Trade and the renewed formation of economic linkages between various cities began a process of sectoral specialization from place to place. Perhaps more importantly, interregional and international trade led to specialization of labour according to different stages of the production and distribution process, i.e. specialization within and between guilds *and* the separation of manufacturing and merchandizing activities.

The implications of such a division of labour were immense. As early as the thirteenth century there were signs of it among Flemish towns, where imported wool from England provided the raw materials for an increased supply of textile goods. Increased consumer demand (brought about by rapid population growth) as well as technological innovations—which made it possible to use unskilled labour to go into real entrepreneurial competition with the weavers' guilds—led to the *putting-out system* and exploitation of peasant wage labour in the countryside. Footloose merchants would provide wool, collect the peasants' cloth for a pittance, and re-sell it dearly elsewhere. This helped monetize the rural economy, weakening the material ties of feudal relations by ending the primacy of the guilds and urban-based merchant capital. It also provided the necessary conditions for emergence of the *separate workplace* or factory, but full development of the factory system had to wait for more fertile ground, in England.

The alliance of town merchants and royal personages was a formidable combination; with steady deterioration of the seignorial economy the royal cause gained increasing momentum. The great riches of empire provided an overwhelming boost in prestige and monetary wealth to the royal–bourgeois alliance, first bringing precious metals to kings and merchants—to pay armies and hire runaway serfs—and then starting the free flow of other raw materials for processing and manufacture. By the beginning of the sixteenth century, France and England were united *kingdoms*, and the 'age of absolutism' followed quickly on their heels: Henry VIII (1509–1547), Elizabeth I (1558–1603), Louis XIV, *le Roi-Soleil* (1643–1715). *This juxtaposition of royal centralism, empire building, merchant wealth, and the beginnings of industrialism was no accident: these events were quintessentially linked.* The landed aristocracy was made subservient to royal power and merchants rose to the prestige of a new peerage, created by fiat.

The monarchy soon found itself at odds with a new alliance of capitalists, however. This faction was formed from the remains of landed wealth, partially renewed by merchant investments in 'property' (Merrington, 1975), and the great commercial and finance brokers, the 'real' power behind the throne. These two groups, aided and abetted by early industrialists, were the makers of the American and French Revolutions and the creators of parliamentary government. Their alliance also outlived its usefulness, though, as 'neo-feudal' country squires found their continuing sources of influence *in an increasingly industrial order* tied to the fortunes of monarchy.[3] Cromwell's Parliaments (1649–1660), the insurgent plantation owners of 1776, and the revolutionary Second Estate of 1789 were very different from the reactionary (royalist) provincials of 1830 and 1848 or the rural pillars of the American Confederacy in 1860. These changes were intimately tied to the expansion of industry and its subsequent capture by finance capital, the roots of which take us to eighteenth century England.

The enclosure movement and putting-out system of manufacture grew up together in the English countryside. They were the foundations of the textile industry which, later, switching from wool to colonial, slave-produced cotton, marked the real beginnings of the 'industrial revolution' (urban-industrialization) in Lancashire. The State set the stage for these transformations through politico-legal action:

> Some 5,000 'enclosures' under private and general Enclosure Acts broke up some six million acres of common fields and common lands from 1760 onwards, transformed them into private holdings, and numerous less formal arrangements supplemented them. The *Poor Law* of 1834 was designed to make life so intolerable for the rural paupers as to force them to migrate to any jobs that offered. And indeed they soon began to do so. In the 1840's several counties were already on the verge of an *absolute* loss of population, and from 1850 land-flight became general (Hobsbawm, 1962, p. 185).

The enclosure movement and its accompanying devastation of rural life was one of the most brutal *rational* acts ever perpetrated by human beings upon their fellows. Whole families were set to wage labour in their miserable cottages. The handloom weavers became the first, rural proto-proletariat, until they too were destroyed by the new urban *factory system*, built on the surplus value extracted from their own labour. Their futile resistance—the Chartist revolt of summer 1842—only brought out the soldiers of the King, sent to protect the steam boilers, power looms, and factory buildings of the industrialists. The age of free industrial capital was born. It subdued the countryside and, after it had ruined the communal guilds of the towns, moved in to create the industrial city. Nearly everyone was forced off the land, leaving it 'free' for the advent of genuine factory-system agriculture. In the cities, human labour had become a salable commodity (almost a free good); capital was king—steam power its queen:

> Devant de tels témoins, O secte progressive,
> Vantez-nous le pouvoir de la locomotive,
> Vantez-nous le vapeur et les chemins de fer.
> (Quoted from R. Picard, *Le Romantisme Social*, by Hobsbawm, 1962, p. 202.)

Urban-industrial development was significantly impacted by other, superficially exogenous developments: (1) the growth of imperial trade and (2) the resultant dominance of finance capital. The industrial revolution was, among other things, a result of the infamous 'triangular' Atlantic slave trade. Blacks from West Africa were ferried across the Atlantic, where they were sold into perpetual bondage to work the Caribbean, Hispanic, and North American plantations. Cotton—planted, harvested, ginned, and baled by slave labour—was then loaded on east-bound ships (also by slave labour) and sent to Liverpool. In Lancashire the cotton was used in dark sweat shops to make textiles, which were, in turn, exported back to Africa and America as well as other European countries and India.[4]

Although the cotton trade was essential to the rise of industrial capitalism in both England and America, it was also responsible for the growth of empire and the eventual dominance of finance capital and the imperial capital cities over industry. As Hymer observed (Hymer and Resnick, 1971), after the 1870s international capital had become a unified force, centred and strategically directed from London. A double role reversal took place during the evolution of industrialization and empire building. The role of the national capitals—i.e. the royal cities—had been that of a merchant prince at the beginning of the eighteenth century; they had created the national markets which provided the economic integration underlying kingly rule. However, as Braudel has argued:

> The obvious fact was that the capital cities would be present at the . . . industrial revolution in the role of spectators. Not London, but Manchester, Birmingham, Leeds, Glasgow and innumerable small proletarian towns launched the new era. It was not even the capital accumulated by eighteenth-century patricians that was to be invested in the new venture. London only turned the movement to its own advantage, by way of money, around 1830. Paris was temporarily touched by the new industry and then released as soon as the real foundations were laid, to the benefit of coal from the north, waterfalls on the Alsace waterways and iron from Lorraine (Braudel, 1973, p. 440).

This world of industrial cities at the turn of the nineteenth century was based on extension of the *factory principle* of organization to the entire fabric of urban social relations and physical development, and then to city/country relationships (Bookchin, 1974, pp. 51–56). Colonial development during the first half of the 1800s, however, consolidated unprecedented wealth and political power in the hands of London financiers. Empire acted as an enlarged national periphery, with resource flows directed towards England and industrial goods sent out to the colonial areas. Despite the importance of manufacturing there was little or no direct investment overseas as we know it today; overall market coordination was done through imperial commercial interests, closely tied to the very centre of government. In effect, the flow of people, capital, resources, and industrial goods was directed from the seat of empire. Even after Britain lost its imperial monopoly, in the last three decades of the century (Hymer and Resnick, 1971), the basic structure of the system continued, with other national variants centred on Paris and Berlin.

Financial interests ran the imperial governments—indeed, became the government—and everything turned in fixed orbits around the imperial cities. National monopolies based on banking and finance were the true royalty, not the King, Emperor, or Kaiser (Lenin, 1917). London and Paris—the nerve centres of the two most far-flung national empires—grew to exceedingly large dimensions by the 1870s. However with the entry of impatient new national competitors, the imperial cities and their expanded national hinterlands were locked on a collision course. Germany's unification in 1871 had established a new, aggressive imperialism in the heart of Europe and Bismarck's imperial designs were founded upon an ultra-modern capitalist industrial base—more than competitive with anything in Britain or France.

At the outbreak of the First World War there were two predominant types of large cities in industrial Europe, the imperial capitals such as London and Paris and the sordid industrial centres like Birmingham, Manchester, Lille, and Frankfurt. Industrial capital and the industrial cities were subordinated to the logic and strategies of the nation-state, dictated and organized from the swollen imperial capital. The countryside was under the rein of the metropolis, as factory relations came to dominate rural as well as urban living. In fact, rural life in the United Kingdom was becoming an endangered species. At the time of the First World War over 75 per cent of the population of the United Kingdom was already classified as urban and metropolitan London counted 6.5 million inhabitants (Hall, 1966, p. 18; 1975, p. 34).

In the two decades after the War this situation was aggravated still further. First-generation industrial technology had placed most industries near essential resource, power, or transport locations. Each region came to be identified with some particular branch of industry, and shared the ups and downs and technological upheavals which impacted that specific activity. In the United Kingdom, where this process had been going on the longest, there was cotton and engineering in Lancashire, coal and iron and steel in South Wales and the North East, and ship building and engineering in Clydeside (Hall, 1975, pp. 83–84). The 1921 census, still reflecting industrial patterns before the Great War, found 60 per cent. of the workforce engaged in three branches of industry: mining and quarrying, metal manufacturing and engineering, and textiles. All three industries produced export goods and two-thirds of their employees were located in England north of the Trent and Scotland (Hall, 1975, p. 84).

The pull of finance interests in the imperial capital, changes in international competition, technology and market preferences, and emergent industrial competition from the more advanced colonial areas ended this nineteenth century arrangement. A new generation of light industry and precision engineering (e.g. electronics, automobiles, pharmaceuticals, foodstuffs, etc.) began to locate in the South and the Midlands—near the money, political power, and concentrating domestic market. After the initial shock of the Great Depression, the 1930s proved a relatively prosperous decade for the London–Birmingham corridor. Continuing depression was the story in Wales, northern England, and Scotland, however:

> Unemployment, 16.8 percent in Great Britain among insured persons in 1934, was 53.5 percent in Bishop Auckland and over 60 percent in parts of Glamorgan; in London it was only 9.6 percent. Despite large-scale migration from the depressed areas—160,000 left South Wales and 130,000 left the North East during the years 1931–9—unemployment rates remained stubbornly high in those areas right through to the outbreak of war (Hall, 1975, p. 87).

It was such conditions in England's Celtic fringe which helped precipitate the first real British move towards regional development planning: The Barlow Commission Report of 1940.

Trends were similar in France, although urban-industrial development was significantly less advanced. A steady movement of people and industry was underway towards Paris, creating in outlying regions what Gravier (1947) was to call the 'French desert'. By 1936 the Paris region came to contain 23 per cent. of the country's industrial population (not including Alsace-Lorraine), compared to 17 per cent. in 1896. Concentration was even more pronounced in the second-generation industrial specializations: mechanical engineering, 30 per cent.; electronics, 49 per cent.; aircraft, 59 per cent.; optical and precision instruments, 69 per cent.; and publishing, 44 per cent. Only declining traditional industries, such as textiles and leather, which were unable to withstand the high wage competition in Paris, remained in the provinces (Gravier, 1972, p. 56). A growing concern over this situation led to still-born regionalization initiatives during the Third Republic (1922) and the Vichy period (1940) (Lagarde, 1977, p. 31).

4. CAPITALIST DEVELOPMENT IN AMERICA

American political unification and urban-industrial development took place under very different historical circumstances. From almost the outset British settlement of North America had been a capitalist venture. The colonies were an integral part of the world's dominant imperial trading system, discussed in the last section. After the War of Independence the merchants and fledgling industrialists of New England and the Mid-Atlantic states joined with southern plantation owners to create one of the world's first federal republics. The United States became, in theory, a separate national economy ruled by its own middle class. Breakthroughs in transport technology and geographic expansion over the next 140 years extended the federal government's sway across the entire continent. Gradual political consolidation of the new territory and formation of a true national market guaranteed Washington's political sovereignty vis-à-vis the individual states.[5]

The same immediate purposiveness could be seen in the growth of America's cities. With the possible exception of New England, where significant changes occurred in the social and spatial form of economic relations between the sixteenth and eighteenth centuries, American cities were meant from the beginning to serve specific mercantile, industrial, and institutional roles in the capitalist economy. Over the course of the nineteenth century there were important changes in the economic and political functions of some cities. Many new urban places grew up from almost nothing, some to become major manufacturing centres, like Pittsburgh, and others, like Chicago, becoming world-class cities (Lubove, 1969; Weber, 1899). As with the country itself, however, urban America did not display the historical overlay of bygone social and economic arrangements common in Britain and France. Cities in the United States were almost exclusively products of contemporary commerce and industry.

Similar generalizations seem valid for the structure of the American economy. While there was always some subsistence agriculture, up through the First World War, even the primary sector had been organized since colonial times on a capitalistic basis (Gunderson, 1976; Hession and Sardy, 1969; North, 1955, 1961, 1966). Although various regions experienced entirely different patterns of economic development—some moving from agriculture to manufacturing, others remaining essentially primary resource producers—all this occurred within the production relations of classic industrial capitalism (Perloff *et al.*, 1960). The only major 'exceptions' were found in the Old South and Hispanic culture areas.

Before the American Civil War slaveholding plantations made an essential contribution to the international development of industrial capitalism, as was noted earlier in this chapter.[6] After 1865, dependent southern development, while making the forced transition to wage labour, continued to maintain certain semi-feudal characteristics as part of the infamous sharecropping system. In Spanish Florida, California, and the Southwest older forms of seignorial economic relations still persisted after a fashion as well. However, as with the early city-regions of New England, these differences, although significant for the history of regional planning (see Chapter 4), were basically idiosyncrasies in the fabric of American capitalism. Dominant patterns were shaped by the marketplace in relation to federal government policy (Cumberland, 1971; North, 1966) and, as such, the United States came to share many traits with the European capitalist economies.

Through the Civil War the main contours of American political and economic life were structured by sectional struggles and alliances between the industrializing Northeast, the export-staple economy of the South, and the foodstuff-producing West. In 1790 the United States had a population of approximately four million, divided almost evenly between the Northeast and the South. Only 200,000 people lived west of the Allegheny mountains and an equal number resided in 'cities', none of them larger than 50,000 in population (US Bureau of the Census, 1960). There was also rough parity between North and South in domestically produced exports, 58 versus 42 per cent. respectively (calculated from figures given in North, 1961, p. 44). Whatever numerical disadvantage the South may have suffered was more than made up for by the strategic importance of cotton. Cotton was the country's largest single export throughout the antebellum period, going from 33 to 58 per cent. of all exports (by value) between 1815 and 1860 (calculated from figures given in North, 1961, p. 233). According to Douglass North, southern cotton provided much of the incentive for northern growth in shipping, finance, and manufacturing, as well as the primary market for western agricultural expansion during most of this period (North, 1961, Chap. 7). Slave-produced cotton was not only 'King' in the South, it was probably the country's leading industry.

The federal union of 1789 had been painfully forged in Philadelphia by a balancing of middle-class interests across the Atlantic seaboard (Beard, 1913). Up through the War of 1812, however, sectional interests were not narrowly identified with one particular economic specialization. The protective tariff of 1816, for example, was primarily the work of two legislators from South Carolina. The South, like New England, hoped for the development of local manufactures, and recognized the need for protection from British competition (Morison, Commager, and Leuchtenburg, 1977). The next several decades witnessed an increasing degree of polarization, however.

The reasons for this were very complex, but several major influences can be identified. At the broadest level, regional specialization in nineteenth century America took place within the context of British imperialism. To an important extent, credit, technology, and manpower were imports from the British Isles, and UK interests and actions played a considerable role in shaping American outcomes. Britain needed an outlet for excess capital, excess goods, and excess labour.[7]

Most of all, perhaps, Britain needed cotton to fuel industrial expansion. Ships sailing from the traditional colonial ports of Philadelphia, Boston, and New York carried cotton to England. British and indigenous capital in the South were invested cumulatively in the large front-end expenditures necessary for slave-produced cotton. As British industrialization marched forward and the beginnings of a national financial and industrial community were forged in the Northeastern United States (through protectionism), the South became ever more firmly locked into a system of producing export staples through chattel slavery. The South's 'peculiar institution' may not have really been so peculiar. Historically it appears to have been a necessary concomitant to British and American industrialization. English and Yankee middlemen also took most of the profits (North, 1961).

The long-term effects were devastating for the South. Public works, which were left to private enterprise and foreign investment by Jacksonian democracy, sprang up almost exclusively to serve the needs of northern finance and commercial interests: e.g. the Erie Canal, the Pennsylvania Railroad. This route over the Appalachians helped link New York trading houses with a growing resource hinterland, and directed European immigrants into the northern Middle-west. Furthermore, the expanding southern plantation system offered little opportunity for absorbing free labour (Morison, Commager, and Leuchtenburg, 1977).

Under the aegis of the protective tariff, American textile and shoe industries grew up in New England and New York, the former based on cheap southern cotton. (Expansion of the plantation system into the 'Black Belt', stretching into eastern Texas, and invention of Whitney's cotton gin, brought the price of cotton down from 31 cents a pound in 1818 to 8 cents a pound in 1831; see Morison, Commager, and Leuchtenburg, 1977, p. 188.) A growing imbalance in regional population south and north of the Mason–Dixon line put the federal House of Representatives in the hands of northerners, and the westward expansion of immigrant-settled free states threatened to do the same with the Senate. The Missouri Compromise (1820) and the Compromise of 1850 prevented this latter event temporarily.

Commerce, industry, and the centre of national population were now located in the North and, importantly, so was the locus of public finance. From 1819 to 1836 the Bank of the United States in Philadelphia directed the nation's money market, ostensibly in favour of eastern merchants and traders. 'Biddle's Bank' was brought down by Andrew Jackson in response to a hue and cry from westerners, only to have its former powers absorbed by the financial community itself: 'Wall Street picked up the pieces of the shattered institution on Chestnut Street, Philadelphia; and a new "money power" in New York soon had more money and power than Nicholas Biddle ever dreamed of' (Morison, Commager, and Leuchtenburg, 1977, p. 192).

Growing abolitionist sentiments in the North were an important element of popular culture, politics, and religion during the 1850s. But the American Civil War (1861–1865) may, perhaps, be more clearly understood in terms of the underlying regional economic rivalries. As early as 1828, the uncompromising tariff system came to be seen as a staggering economic liability by the southern ruling class (Caroll and Noble, 1977). The 'Tariff of Abominations' led to South Carolina's Nullification Act of 1832, which challenged the federal government's sovereignty in creating and enforcing a free national market at the expense of the southern middle class. Andrew Jackson confronted this threat head-on, securing a compromise tariff and defusing the situation for the time being.

Over the next 30 years, however, the balance of population, economic and political power shifted perceptibly northwards. Despite the disproportionate influence of southerners in the national capital, the federal government came increasingly to represent the sectional and class interests of northern finance and northern industry. The free-state/slave-state issue had obviously become a losing battle for the South, threatening an impending takeover of the whole State apparatus by abolitionist forces. Abolition was, among other things, a direct challenge to the very structure of southern economic and social relations. Southerners rightly saw it as revolutionary, in direct opposition to the conservative principles of the constitution of 1789 (Caroll and Noble, 1977, pp. 207–208). When Abraham Lincoln, who opposed the extension of slavery as a matter of policy, was elected to the presidency the crisis came to a head. Led once again by South Carolina, seven cotton-growing southern states had repealed their ratification of the federal constitution and seceded from the federal union by the end of January 1861. Four months later the Civil War began.

The South lost the War, and most historians agree she lost for the same underlying reason she had entered it: her cotton staple economy. The South had no industry to supply its armaments; no railroads to move its troops; few soldiers to fight its battles; and it did not grow enough food to feed itself and its armies. Union blockades stopped the flow of munitions and other manufactures from England, the South's natural ally. Five years of ceaseless fighting bled its manhood dry.

Losing the war destroyed the physical and economic basis of southern society. Much of the existing infrastructure was physically demolished. The brutal exploitation of chattel slavery was ended, and with it the prevailing system of agricultural production. The economic power of the southern ruling class was temporarily surrendered, and northern occupation or 'reconstruction' removed the indigenous middle class from positions of regional political authority. The war also disallowed any possibility of southern influence in Washington for several decades. For most practical purposes, the South was a defeated sovereign State—its economic system destroyed—policed and ruled by a foreign power. Mobilization for war had provided an impetus for accelerated northern industrialization, and after the War the federal government rested exclusively in the hands of the Grand Old Party and eastern capitalists.[8] The stage was set for a continuing polarization of national life.

Narrow economic specializations and glaring class (and racial) exploitation had been the striking features of antebellum southern society. As we have seen, these were particular regional aspects of the broader national and international economic system. Transformation of the postbellum South was also part of a larger picture. Forced land reform in the South, required to come to grips with the realities of wage labour, quickly took the form of the dehumanizing share-cropping system. Land was sold or leased in small parcels to poor whites and newly freed blacks. Lacking even minimal monetary fluidity, advances of supplies from local merchants, backed by northern capital, were needed to provide the wherewithal to live and plant a crop of staple cotton. (Cotton was the only crop that most lenders would allow to be planted.) After a few seasons an increasing number of farmers found themselves unable to repay earlier advances. Each year they went further in debt, consigning their crops to the local 'man' before they were even planted. Within two decades, perhaps the majority of southern farm families of both races were caught in a harrowing cycle of deepening debt. They were not even included in the cash economy, as such. A new regional middle class based on the most extreme forms of usury thus grew up. In such circumstances it soon controlled ever-larger portions of the region's landed property (Beard, 1924). A vast group of landless agricultural workers was created. Some moved westward along the southern tier of states into Texas, only to be followed by the monstrous share-cropping system (Goodwyn, 1978). Others became part of a fledgling urban working class.

Labour-intensive industries, such as textiles and shoes, which could no longer compete in New England, began to spring up in the South, taking advantage of its newly created labour surplus. But in spite of dramatic increases in the number of southern wage earners, low salaries meant meagre income multipliers, while most of the profits left the region, returning to northern investors. By the early 1900s urban-based industrialization was drastically changing the traditional face of the South. Politically the region continued to be disenfranchised, and its traditional culture was being destroyed by machine industry and alien lifestyles (see Friedmann and Weaver, 1979, Chap. 2).

As the South was 'reconstructed' New England also suffered. Its traditional industries began moving southwards and mechanized farming in the Midwest and northern plains totally overshadowed its agricultural potentials. Meanwhile, the federal government, in the hands of the 'captains of industry', took an active role in railroad building and industrial expansion. The 'manufacturing belt' from New York to Chicago became the new national heartland, based on heavy industry manned by European immigrant labour. The plains states were peopled by these same immigrants, whose descendents, however, also soon turned towards the industrializing cities (Perloff *et al.*, 1960).

From 1870 to 1930 dramatic changes took place in the American economy. The forty-eight states were filled in and western resources were finally completely opened to national use. The United States became the world's leading manufacturing country, responding in large part to its own national market. The number of workers in secondary industry surpassed the number in agriculture, and the urban population came to outnumber people living in rural areas (North, 1966). By 1910 the Northeast accounted for 70 per cent. of the country's labour force in manufacturing and 77 per cent. of the value added by secondary industry (Perloff *et al.*, 1960, p. 253). It contained roughly 50 per cent. of the US population, and in New England and the Mid-Atlantic states over 70 per cent. of these people lived in cities (Perloff *et al.*, 1960, pp. 225–227).

With the close of the frontier in 1912 and the unrivalled expansion of industry, American development came to share many of the characteristics already discussed in western Europe. The industrial explosion created a more permanent urban working class, mushrooming tenement districts, and national labour organizations. Rural communities went into severe, long-term decline. The national economy came to be dominated by an urban core which subordinated the rest of the country (Ullman, 1958). The 'Western Empire' in America paralleled in many ways the extended European periphery in Africa and Asia. Until the Great Depression of the 1930s the interests and ideology of government seemed identical to those of big business (Beard, 1924; Tugwell, 1932).

As in Europe, however, living conditions for most people in the burgeoning industrial cities were scandalous, and regional imbalances in productive capacity and incomes became increasingly glaring. Major reform movements during the Progressive Era tried to rectify the worst ills of metropolitan growth through local government reform and urban planning (see Friedmann and Weaver, 1979, Chap. 3). By the 1920s small groups of reformers had also turned their attention to the problem of uneven regional development. Under the extreme conditions of the Great Depression, their concept of 'regional planning' was incorporated into federal government policy as part of the New Deal.

5. URBAN-INDUSTRIALIZATION AND THE UTOPIAN RESPONSE

Urban-industrialization in both Europe and North America had created a new world by the first decades of the twentieth century. Traditional rural communities were in sharp decline, and true national economies were brought into being in France, Britain, and the United States. Urban dominance over the emptying countryside was firmly established by nationwide markets presided over by national political institutions. The metropolis boomed. Regional economic specialization was the order of the day.

This was a painful transition, however. As the economic relations of industrial capitalism restructured society, new classes came into being—notably the urban proletariat—while others began to disappear. Class alliances among changing regional elites brought revolution and civil war. Social movements and revolt exploded in the industrial city and threatened the stability of government. Colonialism and national rivalries led to wars of unprecedented destruction.

It was in response to these momentous events that modern critical thinking and the social sciences first appeared. Utopian socialists, positivists, anarchists, regionalists, and liberal reformers analysed the changing society, fomented revolution, and worked among the industrial poor. Regional planning in its original form was one of their offspring. In the next chapter we will try to untangle its inheritance.

NOTES

1. Urban agriculture, however, was nothing new to the medieval town. As far back as occidental antiquity, even in the most densely populated cities, substantial amounts of land—frequently within the city walls themselves—were set aside for gardening and food production.
2. In large measure, both the beginnings of political regionalism in the late nineteenth century and the contemporary regionalist movement in Europe owe their origins to cultural and spatial 'residuals' of the feudal commune and mercantilist city.
3. Regionalism (or nationalism, if you will) in Northern Ireland and Scotland must be traced historically to the changing political alliances of this period. The rise of the absolute monarchy in England and its later struggles with the ascendant middle class were inexorably linked to English union with Ireland and Scotland. Henry VIII completed the absolutist Tudor State, which his daughter was to rule for almost half a century, with conquest of Ireland in 1542. Under the Stuarts, Ulster was settled in 1610, sealing the 'first' Union of the two countries, and rebellion during the reign of Charles I led to the catastrophic Irish Massacre of 1641. The second conquest of Scotland under Cromwell in 1650 and the formalized first Union in 1652 played an integral role in establishment of bourgeois power during the Interregnum (1649–1660). The Stuart Restoration under Charles II was explicitly engineered through making deals with the Scottish and Irish, which put a king back on the throne of England, in part, by temporarily undoing English hegemony over the two outlying nations.

    The 'bourgeois' monarchy which followed the coronation of William and Mary (1689–1702) was, in fact, the final establishment of middle-class power over 'Great Britain', formed by opening the entirety of the British Isles to unified mercantile capitalist rule with Scottish (1707) and Irish (1801) Union. The concrete circumstances for union were *very* different, of course—Scotland taking part in the spoils of empire and Ireland being one of its first victims—but with the exception of the loss of the American colonies (1783), this marked the beginnings of relentless expansion of the British Empire (e.g. consolidation of India by Hastings, 1780) and something over one hundred years of *'Pax' Britannica*. This, the first unified world system of political economy, was the model of classic capitalist imperialism, based on the London finance market. It provides an important component for explanation of the imperial city's unprecedented growth.

4. India, the jewel of the British Empire, was penetrated in Bengal by Plassey of the East India Company (1757) and then consolidated militarily and politically by Clive and Hastings (1760–1780). She was brought to her knees economically, though, by the forced importation of Lancashire textiles, which ruined a traditional industry that had earlier supplied much of pre-industrial Europe's demand for cloth. Thus, India became underdeveloped. This history—the development of classical capitalist imperialism—makes particularly good reading when juxtaposed with the critical regional theories discussed in Chapter 7. See Panikkar (1959) and Harnetty (1972).

5. Most of the later States had little or no independent political experience before being taken into the Union (except perhaps for the campaign for statehood itself) and were, in large measure, creatures of the federal government in Washington. In fact, New England and the Old South were the only areas which had been settled long enough to have experienced a more organic sort of political and socioeconomic development—a fact of some importance for regional planning after the turn of the twentieth century. Furthermore, in the new western states, the federal government, from the outset, maintained control of all unalienated land and, thus, natural resources.

6. By 1807 over 60 per cent. of the cotton imported into the United Kingdom came from the southern United States. According to North (1961, p. 41), 171,267 of the 282,667 bales of cotton arriving in London, Liverpool, and Glasgow that year came from the United States.

7. Between 1815 and 1860 five million immigrants came to the United States from northwestern Europe: two million from Ireland, over a million and a half from Germany, and three-quarters of a million from the United Kingdom (Morison, Commager, and Leuchtenburg, 1977, p. 211).

8. The GOP (the post-war Republican party) was an explicit civilian–political transformation of the victorious Grand Army of the Republic. Grant, the GAR's commanding general became the republic's first elected GOP president after Lincoln's assassination, and with exception of the Cleveland and Wilson administrations (sixteen years all told) the GOP ruled the country until Franklin Roosevelt's election during the Great Depression in 1932 (i.e. 51 years).

CHAPTER 3

The Precursors of Regional Planning: Utopians, Anarchists, and Geographers

Regional planning initiatives after the First World War came about in response to the transformations of Western society, sketched in Chapter 2. Planners' main themes were drawn from among more general social theories developed during the nineteenth and early twentieth centuries. Unlike planning in the aftermath of the Second World War, however, the first generation of regional planners was clearly influenced, in some measure, by critical interpretations of industrial capitalism.

In this chapter I will summarize the concepts which had a direct impact on the founders of regional planning. In substantive terms, these analyses began as fairly narrowly conceived critiques of industrial organization and working-class morality; later they broadened into a more general (and more radical) indictment of the political and economic basis of the capitalist nation-state. By the end of the nineteenth century, however, the emphasis had shifted once again, this time placing primary importance on local society's relationship with the natural environment. The political ideology of the precursors of regional planning followed a parallel succession over the course of the century, moving from utopianism and radical socialism to reformism as we approach the regional planning movement itself.

A vast body of contemporary literature leaves little doubt as to the deplorable conditions of life and work in the nineteenth century industrial city (see, for example, Addams, 1910; Booth, 1902; Chadwick, 1842; Engels, 1892). Factory workers—frequently young women and children—worked for up to fourteen hours a day on the job. This was mean, degrading work, typically performed under appalling circumstances. Supervision was often brutal and punitive. Pay was uniformly insufficient. Work habits in such an environment were understandably lax and defiant.

Living conditions outside the factory were equally degrading. The intolerable shanties and courts of 'Little Ireland' in Manchester, St Antoine in Paris, or St Giles in London were only surpassed by the tenements of New York (De Forest and Veiller, 1903). Working-class quarters were so overcrowded and

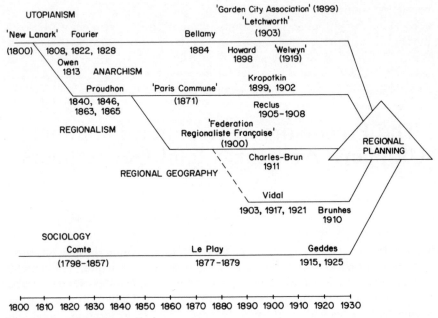

Figure 3.1   Planning precursors

sanitation so poor that infectious disease ran rampant (Engels, 1887; Kellogg, 1914; *Proceedings*, 1910; Sax, 1869). Desperation and drunkeness became a way of life for hundreds of thousands of people.

With public life dominated by such scenes of squalor and despair a new interest in social studies and criticism took root in France and Britain. As shown in Figure 3.1, its most important currents for the history of regional planning were utopian socialism, anarchism, regionalism, regional geography, and regional sociology. (The interconnections between the personages, events, and publications in Figure 3.1 are elaborated during the course of the narrative which follows.) It was the arguments of social scientists and activists concerning political and economic centralization, rural/urban contradictions, class conflict, and environmental relations in industrial society which provided the concepts from which the first doctrine of regional planning would be created.

1. NEW TOWNS: THE IDEA OF STARTING OVER AGAIN

The development of utopian socialism can be traced back at least to the activities of Gracchus Babeuf (b.1760–d.1797) and the prolific writings of Claude Henri Comte de Saint-Simon (b.1760–d.1825) around the end of the eighteenth century (Wilson, 1940). But the important innovators and precursors of regional planning were Charles Fourier (1808, 1822, 1829) and Robert Owen (1813). These two men presented one of the most compelling ideas of nineteenth century social theory: the hope that urban life would be transformed by building new, planned, industrial towns. Despite their widely variant views as to the proper incentives and arrangements for new industrial communities, they both espoused the idea of starting over again *to escape* the prevailing modes of life in the sordid industrial cities. They also agreed on the possibility of cooperation between different classes to bring about a fundamental change in *social* relations, basing their hopes on appeals to the ruling classes and top-down reform.

The utopians choose eclectically from among the rationalist doctrines of the eighteenth century. They believed that 'Truth' would eventually conquer, based on its own self-evident merits. Fourier sat in his room at an appointed hour for years, waiting for convinced capitalists to come forward to finance his schemes. Owen appealed unceasingly for the ruling institutions in Britain to adopt his proposals for resettling the poor. The logical extension of this position was that both men rejected the emphasis on political organization and action which had dominated eighteenth century philosophy and social theorizing.

Because of these convictions—the new town idea, the belief in class co-operation, the faith in rationality, the avoidance of politics—Fourier and Owen were properly called utopian. (The epithet 'socialist' seems more doubtful.) Yet it was these very notions which would be inherited by planners.

Both Fourier and Owen were environmental determinists: they felt that changing the workplace, living environment, and daily routines of factory labourers would change the outcomes of capitalist industrialization. Fourier relied on 'passional attraction' to provide the necessary incentives for social cooperation and control (Goodwin, 1978). Owen believed in the need for proper education and moral example. They both felt that these reforms could be made with only minor changes in existing economic relations, although the actual organization or production would have to undergo fairly extensive modifications.

Owen's idea of the 'cooperative village' was drawn from his own experience in New Lanark, a textile factory-town near Glasgow which he purchased along with his Manchester partners from his father-in-law for £60,000. He managed it as an immensely successful private enterprise from 1800 until he left for the United States in 1824. Home once again in the United Kingdom, Owen became associated with the founding of the British labour movement, as well as the idea of producer and consumer cooperatives. (The latter concept he actually believed to be of little value.) His involvement in these causes helped form the environment in which English Fabianism would take shape, but it was the new town idea—scoffed by industrialists and Fabians alike—that was his direct legacy to regional planning.

Although typically subjected to almost jocular ridicule (or ignored), Fourier probably had the most direct influence on later planning ideas. Both Owenite and Fourierist movements sprang up, founding new communities in rough conformance with their respective ideals, especially in America (see Hayden, 1976). Elements of Fourier's system were passed on directly to other precursors of regional planning, including Pierre-Joseph Proudhon, Elisée Reclus, and Patrick Geddes.

We need not become involved here in the curious systematizing of Fourier's *phalanstery*, nor its strongly hedenistic presuppositions. His contribution to regional planning was the advocacy of largely self-sufficient, free-standing, industrial communities. These were to be based on an association of producers who received shares rather than wages for their labour, and production tasks were to be divided among highly specialized, constantly changing work groups. Fourier placed strong emphasis on mixing traditionally urban and rural pursuits, with a special focus on horticulture and gardening.

Fourier's ideas were passed on to planners via Proudhon and his anarchist followers. Proudhon, a fellow townsman of Fourier, was the publisher's assistant who edited Fourier's *Le Nouveau Monde Industriel et Sociétaire* in 1829, under the supervision of the 'bizarre genius' himself in Besançon. While Proudhon would later rail against most of the tenets of Fourierism, he did adopt Fourier's notion of 'little hordes' doing undesirable community work and also shared the older man's predilection for social analysis through the use of 'series'. More importantly, such anarchist ideas as the rule of purely economic association, freely entered communalism, the labour exchange, and mixing horticulture and industrialism were already prefigured or explicitly developed in *Le Nouveau Monde*. In many ways Fourier's *associative school* seems a direct antecedent to Proudhon's *mutualism*, albeit less analytically developed and, undeniably, more romantic (see Proudhon, 1843; Woodcock, 1972).

Elisée Reclus had been a long-time Fourierist before being converted to Proudhonism and joining the army of the Paris Commune in 1871. Reclus, along with fellow anarchist and geographer Peter Kropotkin, in turn, made significant contributions to the intellectual environment around Patrick Geddes. Through Geddes, the ideas of Comte, Fourier, and Proudhon found their way to Lewis Mumford and the Regional Planning Association of America. (This link is developed more fully below.) In the *Culture of Cities* Mumford (1938, p. 521) was to call *Le Nouveau Monde* 'a book whose wide influence has not yet been adequately estimated or understood'.

The best-known advocate of new town building in the United Kingdom was Ebenezer Howard. Howard's importance as founder of the 'garden city' concept and, by extension, the modern town and country planning movement is richly documented (see, for example, Aldridge, 1979; Cherry, 1974; MacFadyen, 1970; Osborn and Whittick, 1963). Here we are only concerned with his immediate impact on regional planning.

The link between Howard's work and the ideas of the utopians and anarchists is unclear. As yet no one has even attempted to fully analyse his relationship with planning contemporary Patrick Geddes (Boardman, 1978, p. 200). Howard himself attributed his inspiration to Edward Bellamy's utopian vision of Boston in *Looking Backwards* (1884), which Howard read during his sojourn in the United States. In turn, like Fourier, Proudhon, and Geddes, Bellamy had been heavily influenced by Auguste Comte (Hansot, 1974).

However, while in his own work, *Garden Cities of Tomorrow*, Howard (1898, 1902) mentions both Owen and Fourier (p. 123) and cites anarchists Kropotkin (p. 61) and Leo Tolstoy (p. 118), he never explicitly develops any of their ideas and laboriously denies any 'socialist' intentions.[1]

Whatever the origins of Howard's garden city idea, its basic contents and later influence are clear. The garden city was to be a free-standing community of some 30,000 people, located near a larger metropolitan centre. Its fundamental purpose was to change the quality of life for inhabitants of the crowded metropolis and the emptying countryside alike. This was to be accomplished by merging the 'magnetic attraction' of the two environments through the garden city (see Figure 3.2), overcoming the contradictions between town and countryside and making the best attributes of both open to its inhabitants (see Figure 3.3a).

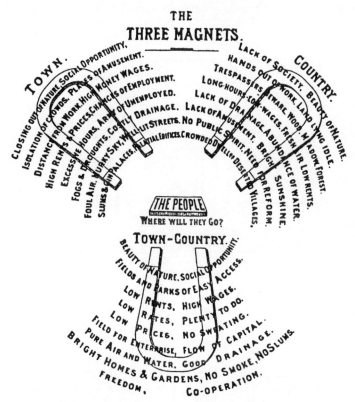

Figure 3.2   Ebenezer Howard's concept of relations between town, countryside, and the garden city (town–country). (Taken from Howard, 1965, p. 46)

Most garden city residents were to be employed in the town itself, while production would be in private hands. Physical planning control and municipal finances would be taken care of by community ownership of the land and a

36

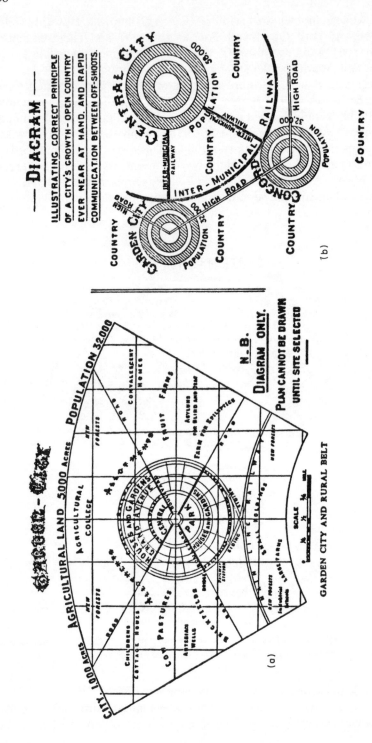

Figure 3.3 (a) Howard's garden city concept and (b) Howard's schematic for regional 'social cities'. (Taken from Howard, 1965, pp. 52 and 143, respectively)

leasehold system. This would be extended to the agricultural strip around the town as well. Farmers were to be private operators, while physical facilities in the town were community property. All increases in rateable value would accrue to the community, amortizing front-end capital investment requirements. The town itself was to stand in a regional system of 'social cities', operating on the same basis (see Figure 3.3b). Informal cooperation and temperance were to be the foundations of social order, not dissimilar from the ideas of Owen, but there would be no taint of the communal enterprise favoured by Owenites. Voluntarism and sound physical planning were the order of the day, and small-scale, decentralized industry was the key to success.

Howard himself was responsible for initiating the construction of two garden cities near London: Letchworth, begun in 1903, and Welwyn, started in 1919. (The plans for these first two garden cities are reproduced in Figures 3.4 and 3.5.) He also helped found the Garden City Association, which later was to evolve into the Town and Country Planning Association. The Association, thanks to the efforts of such men as F.J. Osborn, was largely responsible for initiating the *Barlow Report* (1940) and lobbying for the New Town Act of 1946. Through the influence of Howard's thinking on the Regional Planning Association of America he can also be credited with helping to instigate several new town schemes in the United States during the 1920s, '30s, and '40s (see Stein, 1957).[2]

Over the course of the nineteenth century several important changes took place in the new town idea which must be kept in clear perspective. For Fourier and Owen, utopian communalism was conceived as a method for radically transforming the *social* relations of industrial capitalism. While the new community itself was central to their thinking, it was meant to be a vehicle—a stage—to facilitate more fundamental changes in the mode of human social organization. Although the means of economic production did remain primarily in private hands, formal cooperative institutions and associations played an important role in their thinking. The fact that social conduct would be guided by modified forms of economic production was absolutely essential to their theories.

Without underemphasizing the social content of Howard's garden city concept, it seems apparent, first of all, that small-scale and physical decentralization took on much more tangible importance than they had enjoyed earlier. (On Howard's 'mechanistic' inclinations, see Aldridge, 1979, Chap. 1.) Secondly, except on the question of land rents, a studied effort was made to avoid *any* tampering with basic economic relations. And, third, the garden city itself—rather than the garden city as a nexus of transformed social relationships—became a palpable concern. The utopians' environmental determinism, and especially Fourier's absurd fantasies about design details, should not be allowed to belie their anticipated end result: association, cooperatism. The garden city was first and foremost a pleasant physical environment, despite Lewis Mumford's arguments to the contrary (see Mumford's introduction to Howard, 1965). It was the idea of this physical shell, rather than a model of

38

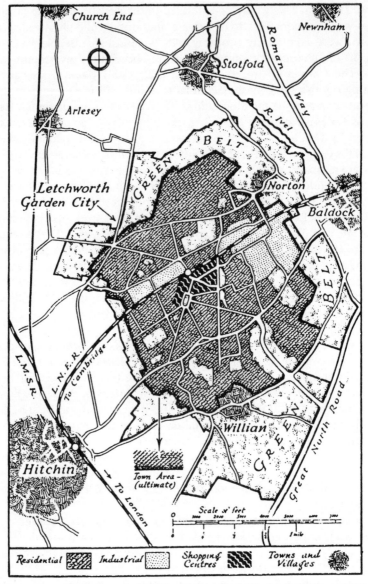

Figure 3.4   Plan of Letchworth (1903). (Taken from Howard, 1965, p. 105)

community, social relations, and economic production, that was Howard's legacy to regional planning. In practice, new town building was to become almost exclusively a physical planning exercise.

2. DECENTRALIZATION, FEDERALISM, AND REGIONALISM

While decentralization was part and parcel of the new town idea, even in

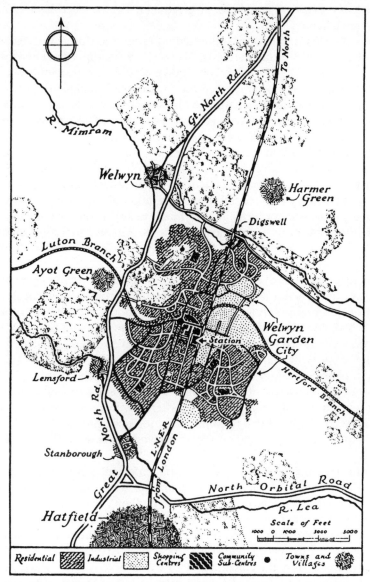

Figure 3.5   Plan of Welwyn (1919). (Taken from Howard, 1965, p. 129)

attaining a doctrinal status it remained basically a practical concern. Proudhon and his heirs, the anarchists, made it a fundamental analytical concept.

All planning histories pay homage to Ebenezer Howard, and most tend to mention the utopian socialists, but, perhaps not surprisingly, the direct links between 'planning' and 'anarchism' have gone unexplored. This is particularly unfortunate, because anarchist concepts of *decentralization of the social economy and regional federalism* prove to have been among the most important influences on early regional planning thought. The connections are clear and easily documented.

As we saw in Chapter 2, urban-industrialization was closely tied, historically, with consolidation of the nation-state and the formation of national markets. Class alliances between the urban middle class and representatives of centralized political authority created the institutional framework necessary for subsequent economic concentration. Industrialization during the late eighteenth and nineteenth centuries led to the creation of a growing yet poverty-stricken urban working class and the concentration of economic activities in a select number of urban locations. With the advance of European imperialism abroad, consolidation of financial power in the national capital, in France and Britain, bolstered these tendencies towards centralization. By the mid-nineteenth century this had led to the formation of a clear-cut national core–periphery structure in which all the dynamic factors in national life were being monopolized by the national heartland: the metropolis and the industrial cities. The outlying provinces suffered from a ruinous cycle of economic dislocation, out-migration, and underdevelopment.

The social theories of Proudhon were a direct reaction to this chain of events in France. As I mentioned earlier, Proudhon was from the town of Besançon, the historic capital of Franche-Comté on the borders of the Jura. Franche-Comté had only been united with France under the reign of Louis XIV in 1678 and was well known for its lingering traditions of strong regional identity and libertarian peasant craftsmen. Proudhon shared directly in these traditions, coming from a peasant family which suffered bitterly from the changing economic circumstances of the period. Proudhon's early critical writings reflect this perspective. Only in his later years, after lengthy residence in Lyons, Paris, and Brussels, did his thinking come round to encompass the impact of large-scale industrial development on the urban working classes (Woodcock, 1962, 1972).

For regional planning there were several important aspects to Proudhonism, the majority important as much for their later omission by planners as their immediate influence. The centrepiece of Proudhon's theorizing was the idea of imminent justice in human society; it formed the springboard for all his other contentions (see especially Proudhon, 1858). For Proudhon this meant that, unhindered by physical coercion or political authority, natural human relationships, springing from the communal nature of everyday tasks, were, per force, rooted in collective effort and equal relations of exchange. This overriding equality was emphatically reinforced by the fact that the means of production—natural resources and technology—were a collective heritage of humankind. More than an individual's own skills and ability, his or her capability to produce their daily needs was based on resources (the free gift of nature), tools and techniques (the collective technological heritage of the past), and cooperation (the work of their fellows). From these considerations Proudhon (1840, 1846) asserted that the most fundamental human relationships were centred around the process of economic production, and production was by its very nature a just, mutual undertaking. Any cumulative alienation of the surplus value of production by particular individuals was a theft of the universal heritage. Thus, he concluded that property, or at least its abuses, was theft.

In Proudhon's view economic and social evolution was a struggle between communism and monopoly. Both were eternal contradictory features of human life. The antinomy could never be destroyed by some higher synthesis; a working balance had to be found.

The centralized political State, which had created the preconditions for nineteenth century urban-industrialization, was, for Proudhon, like its fragmented feudal predecessors, an illicit guarantor of property rights. It was the oppressor of the common people—the power which established the guaranteed monopoly control of the common heritage. There was no practical difference between the absolutism of the *ancien régime* and its various bourgeois successors after the revolution of 1789. The first step toward reestablishing the natural liberty of humankind was, for Proudhon, the dismantlement of all political authority, beginning with the central organs of the nation itself.[3]

In the place of 'industrial feudalism' and central political authority Proudhon argued for *mutualism* and *federation*. Mutualism was a term Proudhon borrowed from the militant textile workers he had mingled and conspired with early in his career in Lyons. It was a concept which for him could balance the eternal contradictions of economic life, without loss of liberty. Under mutualism, individual workers would exercise their absolute right to the means of production (resources and technology), owning their own workplace and tools (Proudhon, 1851). Producers' associations would, thus, inherit the 'great branches of industry' and then join together in *freely contracted* cooperative schemes for vertical and geographic coordination and marketing. These 'social contracts' would do away with all need for central political authority, as the 'natural' (functional) units of society worked out their own freely entered arrangements and government became the mere 'administration of things'. Large-scale tasks of nationwide importance, such as railways and other extensive infrastructure, would be coordinated by similar cooperative agreements.

A central feature of this mutualist scheme was the provision of free credit (Proudhon, 1849, 1851). In order that producers could be guaranteed access to the means of production Proudhon suggested the concept of a *labour bank*. This was based on his acceptance of Ricardo's labour theory of value and the belief that capitalist industrial organization could only be crushed if workers could escape the degrading system of wage labour. The mutual labour bank would make possible exchange based on established units of labour credit, inherent in each producer's work. It would also provide labour-credit loans to meet the expenses of acquiring resource inputs and tools.

Cooperation and mutualist banking established an institutional framework for what Proudhon called the 'social republic'. This larger entity—a free federation—would be formed by a series of constantly re-negotiable social contracts (Proudhon, 1862, 1863, 1865a). Starting at the level of the 'natural unit' (the workplace/living group), contractual agreements would be struck between functional/territorial groupings, beginning with the production unit and the commune, working their way up to the region and finally the European level. Regional units defined by economic production and culture would be the largest building blocks of the social republic. These regions would provide all the prerequisites for a rounded development of human social, economic, and cultural capacities, in an environment free from political and economic coercion.

It is easy to see why contemporary writers have rediscovered in Proudhonism the *locus classicus* of modern theories of integrated self-management (see, for example, Bancal, 1973; Langlois, 1976; Rosanvallon, 1976).

Throughout his life Pierre-Joseph Proudhon declined active leadership of the various political movements which began to draw support from his writing. With his last forced energy, however, in 1864, he wrote what was to become the manifesto of the French working-class movement for the next seventy years. In his *De la Capacité Politique des Classes Ouvrières* (1865b) he set forth an optimistic evaluation of the growing political self-awareness of the French proletariat and specified a three-step process for achievement of the social republic. Proudhon argued that the capacity for political action was founded on possession of three vital qualities: *consciousness, an idea*, and *realization*. Consciousness meant class consciousness, in the sense later popularized by socialist militants. For Proudhon the idea which would serve as the rallying point for working-class consciousness was mutuality. Realization would come about through the struggle for regional federation.

This programme has often been criticized as sheerly utopian, especially within Marxist ranks, but it provided the guidelines for anarchists and anarcho-syndicalists in France, Italy, and Spain up until the Second World War. The influence of Proudhon's ideas, in highly abbreviated form, found their way into regional planning through the work of Peter Kropotkin, Elisée Reclus, and Jean Charles-Brun.

Kropotkin and Reclus were both geographers by profession and anarchists by political conviction. Charles-Brun was a leading later-day Proudhonist and founder of the *Fédération Régionalist Française*. All three men had a direct influence on early regional planning thought which can be traced through their personal associations and the footnotes and bibliographies of the early regional planning literature. The political content and threatening economic implications of Kropotkin's and Reclus' ideas were almost totally ignored, however. This made their schemes more palatable to liberal reformers, but stripped them of whatever practical workability they may have had.

Elisée Reclus met the renegade Proudhonist Michael Bakunin in the socialist circles of Paris during the 1860s. At the time Reclus was a 'Phalanstarian', following the ideas of Charles Fourier. The immediate impact of Bakunin's anarchism on Reclus is not clear, but after Reclus' aborted participation in the Paris Commune of 1871 he became known widely as an anarchist propagandist (Dunbar, 1978). Living in Switzerland in the late 1870s he was a contributor to Peter Kropotkin's radical journal, *Le Révolté*. In his later years Reclus became a sometime colleague of Patrick Geddes, visiting him in Edinburgh (like Kropotkin), teaching a geography course with him during the Edinburgh Summer Meeting of 1895, and falling back on Geddes' aid in attempts to find funding for the 'Great Globe of the World' he wanted to build for the Paris world's fair of 1898 (Boardman, 1978). Geddes is known to have recommended Reclus' work warmly to the attention of his own disciple Lewis Mumford (Dunbar, 1978).[4]

Reclus' contribution to regional planning thought was his anarchist interpretation of the 'social' geography of the world, put down in print between 1905 and 1908 as *L'Homme et la Terre*. *L'Homme* was actually a historical geography of man's relationship with the physical and human environments; it became a political commentary on recent history as it reached the contemporary period. Reclus' interpretation focused on several of the basic tenets of anarchism, attempting to substantiate them with a wealth of empirical detail. He argued that small-scale, collectivist societies had maintained, throughout history, the most fruitful and enduring relationships with the physical environment. Furthermore, these types of societies were demonstrably the most efficient economically and just socially. They established a natural balance with the world they lived in. In a major departure from Proudhonism—which may be attributed to his earlier contact with the Phalanstarians—Reclus also championed an egalitarian, communist distribution of the products of the economy.

The ideas of small-scale communism, cooperation, and natural balance were further developed by Peter Kropotkin, as has been frequently alluded to in the writings of Lewis Mumford. Kropotkin read Proudhon's *System of Economic Contradictions* while doing geographical research in Siberia for the Czarist government, and became converted to anarchism during his first visit to Switzerland in the 1872. He was especially impressed by the disciples of Michael Bakunin (although he never met Bakunin himself) and the anarchist craftsmen he visited during his travels in the Swiss Juras. Later, after having suffered imprisonment at home in Russia, Kropotkin returned to Switzerland, living first in Zurich and then Geneva. In Geneva, where he befriended Reclus and edited *Le Révolté*, Kropotkin became one of the founders of communist anarchism (Woodcock and Avakumovic, 1971).

Peter Kropotkin was a very attractive figure: a kind man, genuinely committed to the well-being and betterment of humanity. His faults as a social theorist were those of the most highminded libertarian. (He always refused to recognize the possible oppressiveness of tradition and the immemorial forms of hierarchical social organization.) During his thirty years of exile in England it was Kropotkin who made anarchism a respectable political idiom. According to George Woodcock (1962), he was the last of the great anarchist theoreticians.

Kropotkin subscribed to all the major themes of anarchism: the economic basis of society, the evils of capitalism and the State, small-scale communal organization, freely entered cooperation, etc. As mentioned above, he substituted a communist distribution of the fruits of labour for Proudhon's system of labour exchange, borrowing Fourier's notion of 'attractive work' as the incentive for enterprise in a libertarian society. Kropotkin's most important contributions to planning were his emphasis on the idea of cooperation or, as he called it, 'mutal aid', and his elaboration of the concept of mixing rural and urban activities to overcome the contradictions created by capitalist urban-industrialization.

Kropotkin's notion of *Mutual Aid* (1902) was that cooperation among members of a species was probably the most important force in natural evolution.[5] His intention was to refute what he considered to be the overemphasis on struggle, competition, and 'survival of the fittest' in the doctrines of Darwin and Huxley. More importantly, he wanted to lay the empirical foundations for an anarchist interpretation of human society as basically a cooperative venture. He quickly moved on from the world of animal evolution to human evolution, and showed that under savagery and barbarism in the medieval city, and even in nineteenth century Europe, mutual aid was the natural basis of human relations and, thus, social organization. It was the imperial State and the artificial capitalist organization of economic production which suppressed, by force, the supremacy of cooperative forms of intercourse.

Importantly, Kropotkin went on to argue that not only was mutual aid the dominant historical form of human social relations but also that it was, in fact, reasserting itself, even under the tyranny of the capitalist State. Far from feeling himself to be a purveyor of utopian experiments, Kropotkin argued that mutual aid was in definite and inevitable ascendance, even being responsible for the very progress of industrialism itself. He wrote:

> To attribute . . . the industrial progress of our century to the war of each against all which it has proclaimed, is to remain like the man who, knowing not the causes of rain, attributes it to the victim he has immolated before his clay idol. For industrial progress, as for each other conquest over nature, mutual aid and close intercourse certainly are, as they have been, much more advantageous than mutual struggle (Kropotkin, 1902, p. 298).

Kropotkin's most enduring contribution to regional planning was made through publication of *Fields, Factories and Workshops* in 1899. Kropotkin argued that the natural trend of secondary industry was geographic decentralization. Furthermore, decentralized production *for local markets* was a desirable rational tendency which should be combined with intensive agriculture—especially horticulture, as envisioned by Fourier. This combination of agriculture and industry—already in progress according to Kropotkin—would overcome the contradictions between town and countryside and allow nations and regions to be virtually self-sufficient in both major sectors of the economy. Small-scale industrial villages, reminiscent of the new towns discussed in the last section, were to be the vehicles of decentralization, providing an environment combining the urban and the rural, as well as an education combining manual and intellectual work. Like institutional economists in the United

States and the Regional Planning Association of America a quarter-century later, Kropotkin felt that new technology, especially electrical power, could provide the necessary material basis for continued economic evolution.

Fields, Factories and Workshops was a technical manual, documenting the possibilities for and realities of decentralized economic production. While it meshed at every point with the political doctrines of anarchism, it took a fairly low-key position on political matters. Its tone and message were extremely attractive to planners and other reformers, and do much to account for Kropotkin's subsequent prestige in such circles.[6] His work played an extremely important role in bringing anarchism out of the conspiratorial underground, but it also must be conceded that his technical proclivities and sweet reason-ableness encouraged the anarchist doctrine to be stripped of its essential political content.

Jean Charles-Brun's work on regionalism brought another central Proud-honist principle to more general international recognition. Like planners' later adaptations of Kropotkin, however, Charles-Brun's apolitical regionalism was a pale reflection of Proudhon's social republic; it had been effectively neutered by fears of official suppression. Both the Regional Planning Association of America and the regional sociologists around Howard Odum at the University of North Carolina were to draw upon Charles-Brun's book, *Le Régionalisme* (1911), for inspiration and verification.

As suggested in earlier chapters, provincial pride-of-place was transformed to regionalism with consolidation of the centralized nation-state and the rise of industrial capitalism. Regionalist sentiments could be progressive or reactionary. Proudhon had incorporated regionalism into his radical socialist formulations as a matter of fundamental doctrine. Regional federalism was the goal to be realized through worker consciousness and struggle.

Regionalism appeared in France, according to many authors, after the brief flowering of the Paris Commune of 1871 (Flory, 1966; Mayeur, 1977; Vigier, 1977). The ideology of the Commune had been heavily influenced by Proud-honist ideas, and regional autonomy was one of its central tenets (Andrieu, 1971; Bancal, 1971). Over the next three decades, however, the regionalist movement came, purposely, to encompass a wide range of advocates, from federalists to provincial intellectuals interested in 'local colour' (Vigier, 1977). This was reflected in the composition of the *Fédération Régionaliste Française* and its journal, edited by Charles-Brun, *L'Action Régionaliste*. Regionalism became a muted protect, 'affirm[ing] the recognition and respect of [regional] diversities, in the same fashion as the respect and indissolvability of French unity' (Flory, 1966, p. 111).[7]

According to Flory, the Federation's programme was as follows:

1. FROM AN ADMINISTRATIVE STANDPOINT
   a) Division of France into homogeneous regions;
   b) creation of regional centres;
   c) rational decentralization, management of local affairs locally, regional affairs at the regional level, and national affairs by the State;

d)　creation of a jurisdictional body charged with arbitrating conflicts between individuals, localities, regions and the State.

2. FROM AN ECONOMIC STANDPOINT
a)　Freedom of regional and local initiatives;
b)　reconciling the economic interests of each region;
c)　use of the almost inexhaustible resources which can furnish to a renaissance a broad regional life and utilization of the intelligence, virtue, and energies with which our provinces overflow.

3. FROM AN INTELLECTUAL STANDPOINT
a)　Tailoring of the three-level educational system to regional and local needs;
b)　development of private initiatives in the realm of humanities, science and art.

Its goals and means are and remain:
1. to renovate political, economic, literary, artistic and scientific regional life.
2. By an intelligent choice of traditions and by teaching local history and folklore, foster in children pride of ancestry and birthplace, rooting patriotism in tangible realities.
3. Bring pride of profession to the worker and peasant.
4. Give to everyone, liberated from the suffocating grip of the State, the taste of initiative.
5. Bring together all the groups and key individuals interested in the regionalist cause.
6. With their help, organize locally and in Paris press campaigns and lectures for the propagation of regionalist ideas and the protection of local interests (Flory, 1966, p. 111; my translation).

This programme, including its patriotism, moralism, and emphasis on education, anticipates almost precisely the position to be taken by regional planners during the 1920s and '30s.

3. THE REGIONAL METHOD

At the turn of the century regionalism was also conceived as a mode of intellectual enquiry. Although it goes unmentioned in the standard intellectual histories (e.g. Freeman, 1961; Hartshorne, 1939; James, 1972), regional sociology and regional geography seem to have been another extenuated response to political and economic centralization. They transformed the programmes of regionalists like Charles-Brun into an academic paradigm.[8] Planners such as Geddes played an important role in its propagation, and regional planning as a field of professional activity leaned heavily on the new 'regional sciences' for its analytic content. The two most important influences were the regional sociology of Le Play and Geddes and the French school of regional geography, founded by Paul Vidal de la Blache. It is convenient to begin with the latter.

Regional studies, as well as popular regionalism, took root in the first decades of the French Third Republic (1870–1940). Unlike Reclus' geography, which was influenced early on by German thinking (Reclus was trained at the University of Berlin, where he had contact with the great German geographer Carl Ritter), Paul Vidal de la Blache seems to have been a true child of France. Vidal studied at the *Ecole Normale Supérieure* and the French

School of Archeology in Athens.[9] In 1898 he was appointed to the chair of geography at the Sorbonne. While teaching in Paris Vidal founded the seminal organ of French academic geography, *Annales de Géographie*, and trained the generation of scholars—Gallois, Brunhes, Blanchard, Demangeon, de Martonne—who were to establish geography as a subject of study throughout the French university system. For all intents and purposes, French geography, almost to the present day, is synonymous with *la tradition vidalienne*.

The object of study for Vidal and his followers was the 'geographic region' or *pays*: a complex locale—the creation of interaction between human culture, social institutions, technology, and the natural environment. In practice, the 'regions' studied by French geographers were most typically the historical market areas of mercantile trading cities and the feudal principalities and dukedoms which pre-dated and contextuated the modern French nation, as discussed in Chapter 2 (Blanchard, 1906; Demangeon, 1905; de Martonne, 1902; Gallois, 1908; Levainville, 1909; Sion, 1908; Sorre, 1913; Vallaux, 1906; Vidal, 1903, 1917). These were recognized geographic areas, with traditional boundaries, economic specializations, settlement patterns and housing types, and, frequently, a residual local dialect or language. The physical artifacts of human occupance were fashioned over the centuries by cumulative human choices generated as a cultural response to the many possibilities laid before society by the natural environment. This bundle of cultural choices Vidal referred to as the *genre de vie*, and the reciprocal relationship formed between humankind and the physical world Vidal called *possibilism*.

The regional unit, comprehending several thousand square kilometres, perhaps, was the primary contact point between human beings and nature. Marked out, as it was, before the rise of the nation-state and industrial capitalism, it formed the 'natural' boundaries of human social economy, political dominion, and cultural habitudes. It was the varigated yet human-scale environment in which the potentials of development were worked out, creating the rich mosaic of medieval France: Brittany, Flandres, Aquitanine, Auvergne, Alsace-Lorraine, Picardie, Gascony, Normany, Languedoc, Provence. It was this variety which allowed the exercise of human creativity through direct participation in all aspects of life. This active, experienced environment was the motor force of human development; the almost sensual reciprocity between men and women and their surroundings was the seat of comprehensible liberty and the mainspring of cultural evolution. The centralized nation-state and its attendant machine industry were suppressing and submerging all this. To save the regional habitat it would be necessary to study and understand it according to Vidal, and the way to understanding was through indepth, holistic surveys of the regional environment. This led to a series of 'regional monographs', the classics of which are cited above. They provided a fine-grained, detailed knowledge of all dimensions of regional life, as well as a historical analysis of how a particular *paysage* or landscape was formed by a specific *genre de vie*. Systematic summaries of the French geographers' methods and findings were set out in textual form by Brunhes (1910) and Vidal (1921) and exercised a considerable influence on planners in both Britain and the United States (Mumford, 1938; Odum and Moore, 1938).

While French geography lent academic legitimacy and literary eloquence to the regionalist paradigm, it was an earlier French engineer and sociologist, Frédéric Le Play (b.1806–d.1882), as interpreted by Patrick Geddes, who presented planners with the clearest outline of the regional survey method. Geddes first learned of Le Play's work through a lecture given by Edmond Demolins, one of Le Play's disciples, at the Paris World Fair of 1878. Le Play's notion of an empirical, regional sociology became one of the guiding concepts for all of Geddes' later social thought.

Le Play, a retired professor of metallurgy at the *Ecole des Mines*, believed that the social theories of Comte and radical propagantists such as Proudhon were caste in much too grandiose terms. Abstract notions such as 'class' masked more than they revealed (Brooke, 1970). Le Play also believed in an economic interpretation of history, but felt that technical progress and war between the various European nation-states were the most obvious determinants of contemporary life. His suggested cure for the depredations of progress was the achievement of *social peace*, through a regeneration of regional life (Le Play, n.d.) based on a thorough understanding of the reciprocal relationship between society and its local environment.

Working towards this latter goal he argued that the proper objective of sociology was to promote the well-being of the family, the fundamental unit of social organization. The family could only be understood and rendered aid through application of the methods of empirical science. To this end he set out on innumerable journeys across France, collecting huge amounts of empirical data about the country's thousands of working-class communities (Le Play, 1877–1879).

Le Play attempted to analyse this information through a three-part conceptual framework: *famille, travail, lieu* (i.e. folk, work, place). He argued that the well-being of the family was a product of the work it was engaged in, and that this, in turn, was impacted by its place of residence. In more primitive circumstances, place—the geographical environment—determined in large measure the family's occupation. At a higher level of technology this became an evidently reciprocal relationship.

It was this trilogy of folk–work–place which attracted Geddes. In it he saw a parallel to the basic triad of biology: organism, function, and environment (Stalley, 1972, p. 10). He was compelled by the idea that such a framework provided a tool for organizing and interpreting regional social surveys.

It is difficult to overemphasize the contribution of Patrick Geddes to the later development of regional planning. His rambling analysis of urban life roundly condemned the dirty sprawling industrial cities, or 'conurbations', which he attributed to an outdated 'paleotechnic' order (Geddes, 1915).[10] Originally trained as a biologist, Geddes has been variously identified as a sociologist, planner and geographer. He was, indeed, a man of catholic interests, and came in contact with many of the leading social critics and reformers of his day:

Elisée Reclus, Peter Kropotkin, John Dewey, Thorstein Veblen, and Jane Addams, to name only a few of the most prominent.[11] He founded the Outlook Tower in Edinburgh (a curious 'sociological laboratory'), the Edinburgh Summer Meetings, which drew a distinguished list of visitors from all around the world, and the *Collège des Ecossais* in Montpellier, France. He was also associated with organization of the Sociological Society and Le Play House in London, and suggested the formation of a National Geographic Institute in Britain and the implementation of regional geographic surveys for the entire country. Suffice it to say that Geddes was an extremely active advocate of planning and planning-related causes, and that during his professional life he carried his ideas as far afield as Chicago and India (see Boardman, 1944, 1978; Kitchen, 1975). Lewis Mumford was probably Geddes' most famous understudy (in the field of planning) and through Mumford, despite his own aversion to writing, Geddes exercised a significant influence on the evolution of regional planning (see Mumford, 1966).

Patrick Geddes was a life-long francophile, and some of his most important inspirations came directly from the ideas of Auguste Comte and Frédéric Le Play. From Comte, Geddes drew his interest in the classification of the sciences and application of the scientific method to the study of society, especially regional society.[12] As we have already seen above, Le Play also provided Geddes with the fundaments of his regional survey method.

Geddes began by transposing Le Play's *famille, travail, lieu* into place–work– folk, typically placing emphasis on the physical environment and how it related to human occupations. This formed the basis for the best known of Geddes' notorious 'thinking machines', reproduced in Figure 3.6. Place–work–folk cuts a diagonal across the diagram and then provides a set of interrelated categories for the concrete study of human settlements. For example, starting in the upper lefthand corner of Figure 3.6, the study of a particular place (i.e. geography) can be extended across the first row of the diagram to focus on the particular attributes of place–WORK (or geographical economics) and, finally, to place– FOLK (or geographical anthropology). In similar fashion, going down the first column, geography yields to WORK–place (or economic geography) and then to folk–PLACE (or anthropological geography).

To the uninitiated this ninefold breakdown of topics for study seems a bit pointless. There is no apparent causal chain running through the various subcategories, and Geddes was never able to explain their systematic interconnections very satisfactorily (see Geddes, 1915). For Geddes, however, they provided the components of what he called the 'valley section' or 'valley plan of civilization', which was to be his model for conducting regional surveys and understanding regional society (see Geddes, 1925). The 'valley plan' is reproduced in Figure 3.7.

This diagram was meant by Geddes to represent a hypothetical river valley running from the mountains to the sea. Its dynamic focus was the coastal city in the far right-hand corner and its purpose was to demonstrate the essential 'unity in diversity' of the entire region. Geddes argued that each area within the

PLACE *(Geography)*	Place–WORK (or geographical ECONOMICS	Place–FOLK (or geographical ANTHROPOLOGY
WORK–place (economic GEOGRAPHY	WORK *(Economics)*	Work–FOLK (economic ANTHROPOLOGY
Folk–PLACE (anthropological GEOGRAPHY	Folk–WORK (anthropological ECONOMICS	FOLK *(Anthropology)*

Figure 3.6   Geddes' 'place–work–folk' thinking machine. (Taken from Stalley, 1972, p. 283. Originally reproduced in Boardman, 1944)

Miner   Woodman   Hunter   Shepherd   Peasant   Gardener   Fisher

Figure 3.7   Geddes' 'valley plan of civilization' or 'valley section'. (Taken from Stalley, 1972, p. 322. Originally reproduced in Mairet, 1957)

region, based on its natural environment and relationship with the city, would be the site of a particular 'nature–occupation' (or place–WORK). The conditions of its residents, their culture, and their aptitudes, would be related to this particular occupation. Conversely, the regional city would be impacted by the degree of development of all its various contributing localities. Thus to start the task of replanning the industrial city, it was first necessary to have an indepth knowledge of the nuances and special attributes of not only the city itself but its surrounding region as well. Thus Geddes arrived at the idea of the regional survey.

On this point, Geddes carried Le Play's empiricism yet a step further. Not only should would-be reformers make a reconnaissance survey of the geographical area in question but both planners and the general citizenry should literally walk every mile of the region, keeping in mind the model of the 'valley plan of civilization' and immersing themselves in the concrete realities of regional life. From this vantage point a specific knowledge of the problems and potentials of a particular city could be thoroughly absorbed, and the task of reconstruction could then be undertaken.

This was the logic behind Geddes' call for complete regional social surveys of the United Kingdom, a task finally accomplished under the leadership of L. Dudley Stamp in 1947. Despite its apparent conceptual confusion the Valley Section Survey was to hold a special appeal for planners. Its influence can be traced not only in the British Land Utilization Survey but also in the work of influential members of the Regional Planning Association of America (e.g. MacKaye, 1928; Mumford, 1925, 1938). The idea of the river basin as an appropriate area for planning, although largely influenced by unrelated historical factors in the United States, would play an important role in the subsequent development of regional planning.

4. SUMMARY: REGIONAL PLANNING THEMES

The precursors of regional planning worked and wrote over a period of approximately one hundred years, from the early part of the nineteenth century to the first decades of the twentieth century. Nearly all of them found conditions of life in the burgeoning industrial cities deplorable, and most proposed schemes by which they might be ameliorated. In Table 3.1 I have attempted to summarize their central themes. Fourier, Owen, and Howard set out proposals for escaping the dehumanizing environment of the cities through the founding of new industrial communities. Proudhon, Reclus, and Kropotkin argued for dismantling the capitalist economy altogether and doing away with the authoritarian central State. They proposed instead a self-managing social economy with decision-making power devolved to the local and regional levels. French regional activists and geographers carried on the Proudhonist tradition, shorn of its radical political content, making the region a primary focus for cooperation, education, and academic research. The regional sociology of Le Play and Geddes aimed at improving the life of the industrial working class through obtaining detailed knowledge of conditions in different urban regions and using this information as the basis for fundamental self-improvement.

As we have seen, most of these writers were informed by their predecessors' thinking, and it is possible to trace a number of explicit themes from Fourier's time down to the work of Kropotkin and Geddes. Most important among these common threads were probably: (1) a basic revulsion with the industrial city, (2) a strong negative reaction to economic and political centralization, (3) the conviction that regional life and culture in the outlying provinces must be restored and that this could be accomplished through (4) a mixing of rural and urban occupations (to overcome the increasing contradictions between town and countryside) and (5) a combination of manual and intellectual tasks, beginning at the essential level of education. Most of these men also shared a certain nostalgia for the rustic life and felt that a significant change was necessary in relations between industrial society and the natural environment. Kropotkin and Geddes, for example, believed that all these goals were part and parcel of the same fundamental problematic.

Inspecting Table 3.1 more carefully, however, it becomes apparent that

Table 3.1 Important themes of the direct precursors of regional planning

	Attractive work	Mutualism/ cooperation	Emphasis on education	Moral improvement	Improving working conditions	Improve management	Communist distribution	Regionalism	Regionalization	Regional survey	Improving environmental conditions	Reinforcing the family	Preserving regional culture
The utopians													
Charles Fourier (French, 1772–1837)	X	X	?		X	?					X	X	
Robert Owen (Welsh, 1771–1858)		X	X	X	X	X						X	
Ebenzer Howard (English, 1850–1928)		?	X	X								X	
The anarchists													
Pierre-Joseph Proudhon (French, 1809–1865)	X	X		X	X			X				X	
Elisee Reclus* (French, 1803–1905)		X			X		X	X			X		X
Peter Kropotkin* (Russian, 1842–1921)	X	X	X	X	X	X	X	X			X		X

The regionalists
Jean Charles-Brun
(French, ?)

The regional
geographers
Paul Vidal de la
Blanche
(French,
1845–1918)
Jean Brunhes
(French,
1869, 1930)

The sociologists
Auguste Comte
(French,
1798–1857)
Frederic Le
Play (French,
1806–1882)
Patrick Geddes
(Scottish,
1854–1932)

* Reclus and Kropotkin both enjoyed well-established reputations as professional geographers.

there were a number of important substantive and ideological differences between the various writers. Despite their calls for reform, the utopians and the regional geographers and sociologists were basically political conservatives— and so was the *Fédération Régionaliste Française*. Their proposals posed little threat to the existing order of economic and political power in industrial society. To an important extent, they were also environmental determinists. Their theories of the depredations of urban-industrialization were closely linked to fairly reified conceptions of changing technological relationships between society and the natural environment.

Among the precursors of regional planning the most complete analysis of the social and economic basis of working-class living conditions and changing environmental relations was presented by the anarchists—Pierre-Joseph Proudhon and his followers. They also set out the clearest programme of social reforms—often revolutionary in character—which would be necessary if the core/periphery structure of metropolitan dominance and regional decline was to be reversed. These solutions struck at the very heart of the historical trends outlined in Chapter 2: the centralizing nation-state, the consolidation of economic and political power within the capitalist classes of society, and the subsequent drain of all national energies into the metropolitan capitals and industrial cities. There was an obvious understanding on the part of anarchist theorists that functional economic power would have to be reestablished within the hands of territorial communities if the disparities between social classes and different geographic regions was to be overcome.

This brings us back to an observation I made earlier concerning the influence of Proudhon on the evolution of regional planning. As we have seen in the preceding pages, many of the elements of Proudhonist doctrine found their way into the thinking of men closely associated with the emergence of regional planning. However, the more radical tenets of libertarian theory were filtered out in the process. In the next chapter, recounting the theories proposed and actual projects undertaken by regional planners in the decades between the two world wars, it will become evident that losing sight of the ties between institutionalized economic power and community development played a major role in shaping the course of regional planning.

NOTES

1. This statement needs to be modified slightly, perhaps. Howard was willing to allow the term 'socialism' if it were defined as meaning: 'a condition of life in which the well-being of the community is safeguarded, and in which the collective spirit is manifested by a wide extension of the area of municipal work' (Howard, 1898, p. 131). But this definition was little more than a sleight of hand on Howard's part, because it was offered in the context of a lengthy anti-socialist argument. For an alternative view, see Aldridge (1979, Chap. 1).
2. The early Howard-inspired new towns in the United States include the work of Clarence Stein and Henry Wright, as well as the new communities first built by the Tennessee Valley Authority. According to Rexford Tugwell, his Green Belt New Towns had rather different origins (see Friedmann and Weaver, 1979, Chap. 3).

3. Proudhon's interpretation of the nature of economic contradictions and his complete rejection of the use of centralized political power were the substantive bases for his break with Marx. These fundamental theoretical differences continue to define the irreconcilable breach between libertarian and authoritarian socialists. For the classic Marxist perspective, see Marx (1847, 1875), Engels (1872, 1873, 1880), and Marx, Engels, and Lenin (1972). The best synthesis of Proudhon's thinking is given in Voyene (1973).

4. Geddes was intimately tied up with the Reclus family. Elie, Elisée's sociologist brother, was also Geddes' guest in Edinburgh, and Paul Reclus (Elie's son) collaborated with Geddes in his final venture, Le Collège des Ecossais. Paul became Geddes' executor in France after his death (see Boardman, 1944, 1978; Kitchen, 1975).

5. Mutual Aid shared this in common with Patrick Geddes' interpretation in The Evolution of Sex (Geddes and Thompson, 1889), written three years after Geddes and Kropotkin first met. There is no indication in Mutual Aid that Kropotkin was in any way guided by Geddes' earlier work, but their similarity in outlook undoubtedly accounts in part for the warm welcome which some of Kropotkin's other ideas received in Geddes' circle of influence.

6. The book's theme and appeal are summarized in the following passage:

> . . .while all the benefits of a temporary division of labour must be maintained, it is high time to claim those of the *integration of labour*. Political economy has hitherto insisted chiefly upon *division*. We proclaim *integration*; and we maintain that the ideal of society—that is, that state towards which society is already marching—is a society of integrated, combined labour. A society where each individual is a producer of both manual and intellectual work; where each able-bodied human being is a worker, and where each worker works both in the field and the industrial workshop; where every aggregation of individuals, large enough to dispose of a certain variety of natural resources—it may be a nation, or rather a region—produces and itself consumes most of its own agricultural and manufactured produce. (Kropotkin, 1899, pp. 26–27)

7. This and the following quotation from Flory were taken from Lagarde (1977, pp. 27–28).

8. Given the chronology of regionalist origins presented in this chapter, it is most striking that so little has been made of the apparent links between regionalist movements and scholastic regional studies. The ideas of Vidal de la Blache in France, for instance, reflect quite faithfully the contemporary arguments of the *Féderation Régionaliste Française*, including his emphasis on the identification of homogeneous regions and regional centres.

9. It should be noted that in later years (1899), Vidal did draw upon the ideas of Friedrich Ratzel in forming his concept of 'possibilism' (i.e. the parameters of human action are very broadly bounded by the possibilities offered by the physical environment, but the choice among myriad alternative paths of development is a product of human culture and intelligence).

10. Geddes couched his ideas in frequently incomprehensible vocabulary, coining such terms as *conurbation, eutopia, paleotechnic, neotechnic,* and *geotechnics.* Conurbation, the only concept to find its way into more general usage, signified the consolidation of several nearby industrial cities in a chaotic metropolitan development. Eutopia referred to a realizable utopia. Paleotechnic and neotechnic were used, respectively, to identify the sociotechnical order of industrial capitalism and the more humanized environmentally sound technical culture Geddes hoped would succeed it. Geotechnics was a synonym for ecology or human geography—what Geddes felt would be the extension of his regional survey method.

11. Like most of the precursors of regional planning, Geddes was an outspoken critic of Marx. His own thinking was extraordinarily eclectic and confused, however. For a

discussion of the influence of John Dewey and Thorstein Veblen on regional planning, see Friedmann and Weaver (1979, Chap. 2).

12. The idea of applying the scientific method to society dates back at least to Giambattista Vico and his *La Scienza Nuova* (1744). While Vico's 'new science' was rediscovered in the nineteenth and early twentieth centuries by such thinkers as Georges Sorel and Benedetto Croce (see Hughes, 1958), it was the formulations of August Comte which influenced regional planning. Proudhon owed a seldom acknowledged debt to Comte, which he passed on to later anarchists like Kropotkin (Woodcock, 1962), and Geddes was much affected by Comtian ideas, being a declared 'positivist' much of his life (Boardman, 1978).

The importance of this connection is that Comte's view of science and its application to society was an extremely primitive one, and his arguments were almost immediately rejected in most circles because of his excesses (Comte, 1852, 1969, 1972). In his 'classification of the sciences', much mimicked by Geddes, he claimed to be able to achieve the same degree of rationality and objectivity in sociology that could be attained in the physical and biological sciences. Indeed, he argued that sociology should be considered the queen of the scientific disciplines. His later attempts to use sociology as the basis for religious observances largely discredited the entire *corpus* of his work. Le Play, for example, would not use the term 'sociology' coined by Comte for fear of being associated with Comtian follies and foibles.

CHAPTER 4

Comprehensive River Basin Development

As we saw in Chapter 3, most of the influential precursors of regional planning lived and worked for important segments of their careers in either France or Britain. Their ideas, which were to play a major role in early regional planning, reflected the important social and economic changes which had taken place in those countries during the consolidation of State power and the rise of industrial capitalism—roughly the eighteenth and nineteenth centuries. Perhaps surprisingly, however, little progress was made in western Europe before the Second World War towards formulating this body of concepts into a doctrine applicable by reform-minded politicians to subnational problems such as localized unemployment and regional inequalities in living standards.[1]

It was in the United States during the 1920s that regional planning first made its appearance, becoming one of the last outliers of the Progressive reform movement. Strongly influenced by the work of Patrick Geddes, the Regional Planning Association of America (RPAA) was founded in New York City in 1923. The RPAA was a small group of architects, housing reformers, institutional economists, urbanists, and foresters, including among others Clarence Stein, Lewis Mumford, and Benton MacKaye. They were responsible for several influential publications which welded many of the fundamental components discussed in the last chapter into a comprehensible synthesis. The Association's members were also involved in a number of practical planning projects during the late 1920s and '30s, but it was their role as propagandists which proved the most important. They disseminated their arguments in America so successfully that by the fourth year of the Great Depression, when Franklin Roosevelt was elected to the presidency, regional planning became one of the earliest anti-depression planks of the New Deal. With publication of Lewis Mumford's famous account of *The Culture of Cities* (1938) the American codification and interpretation of regional planning found its way back across the Atlantic to Europe.[2]

The RPAA was not the only group in America to discover and advance the cause of regional planning in the years between the two world wars. Equally important were several regional sociologists at the University of North Carolina at Chapel Hill, working under the leadership of Howard W. Odum.

These members of the 'New South' Movement were part of a more general reaction against the exploitation and underdevelopment of the Southeastern States described in Chapter 2. They were cultural regionalists, constrained by historical circumstances to work for reform within the confines of the national state. Not surprisingly, Odum and his followers were directly influenced by Charles-Brun and the French regionalist movement. Odum's approach to 'regional–national social planning' was strikingly similar to the platform of the *Fédération Régionaliste Française* discussed in Chapter 3. The essential difference appears to have been Odum's explicit emphasis on the *rational, scientific* nature of regional planning, and the receptiveness of the American political culture to such an appeal after several years of depression.

1. REGIONAL PLANNING: THE THEORISTS

The members of the Regional Planning Association of America worked primarily on the northeastern seaboard of the United States, the most highly developed section of the manufacturing belt which had arisen in the decades following the Civil War. Urban-industrial development was in full swing in southern New England and the Mid-Atlantic states, and the problematique of regional planning, as conceived by the RPAA, was to stop the 'flood of metropolitanization' and the growth of 'dinosaur cities' which were overwhelming earlier patterns of human settlement and rural/urban relations (Mumford, 1925a). Like Geddes, who visited with Mumford and lodged at the New School for Social Research in New York during the summer of 1923, the RPAA had an evil foreboding about the spread of 'conurbations' and the destruction of apparently stable human/environmental relationships by the 'paleotechnics' of capitalist industrialism (Mumford, 1966; Sussman, 1976, Chap. 1). Most importantly, taking their key from Oswald Spengler, the association believed that the 'third migration', draining all social life and activities into the cancerous metropolis, must eventually spell the decline of Western civilization, which, historically, had been founded on a balance between cities and their surrounding regions and an active, decentralized cultural life (MacKaye, 1928; MacKaye and Mumford, 1929; Mumford, 1925a, 1938).[3]

It would be disingenuous at this point, perhaps, to fault the RPAA for the breadth of its vision and the grand brush strokes with which it painted the plight of industrial society, especially when one takes into account the erudite, voluminous writings of Lewis Mumford and their continuing impact on international opinion (see Johnson, 1983). It seems unavoidable, however, to compare the RPAA's analytical position and social concerns with their immediate predecessors, discussed in the last chapter. From this perspective their most influential publications, especially MacKaye's *The New Exploration* (1928) and Mumford's *The Culture of Cities* (1938), appear somewhat stilted and vague. Only the lesser French geographers and the confused flamboyance of Patrick Geddes' writings leave the reader with such a complicated, muddled image. While this was partially a matter of style, there is undeniably an important substantive element to it as well. While tangible things such as workers' housing, big city slums, and industrial centralization are objects of discussion, the debate typically unfolds on a much more abstract level, and the terms of reference are frequently drawn from esthetics, higher criticism, and a kind of environmental transcendentalism.[4] The end result is not dissimilar from the feeling conveyed by much of the environmentalist writing of the early 1970s, with which most contemporary planners are familiar. There is a lingering impression that one had entered a fanciful world, a world whose shibboleths and romantic images are only tenuously related to the people, events, and circumstances it set out to describe and analyse.

The problems and solutions envisioned by the Southern Regionalists were more straightforward and earthbound, if less profound. Three quarters of a century of political disenfranchisement and economic underdevelopment—on top of the earlier socioeconomic pathology of chattel slavery—had left the South's economy in an unenviable state of structural undevelopment and outside dependence. Agriculture remained largely a matter of export-crop production by means of semi-identured smallholders and tenants. Urban-industrialization after the turn of the twentieth century took the form of an enclave economy, exploiting surplus labour forced off the land by the share-cropping system to produce wage-goods for export, such as textiles and shoes. This now classical pattern, reproduced throughout much of the dependent world periphery in the later twentieth century, exacerbated internal class relations and helped produce a marked cultural retrogression and meanness, tinged increasingly by alien elements. Imported urban-industrial values and lifestyles were being overlain on a base of reactionary sectionalism.

The Southern Regionalists proposed to reverse these predominant trends, to restore a higher degree of equality between different parts of the United States, through a programme of relatively autonomous institution building, education, and resource development at the regional level (Odum, 1934, 1936; Vance, 1935). Their strategy of 'regional reconstruction' was framed in the context of the American federal system and was suggested as an approach applicable to all parts of the country, but their interest and concern were clearly stirred by a passionate commitment to rebuilding the South itself (Odum, 1936; Odum and Jocher, 1945; Odum and Moore, 1938). It is striking how the southerner's analysis anticipated, in substance if not in tone, the contemporary work of critical theorists like Andre Gunder Frank (see Chapter 7); whole paragraphs and pages could be transposed from one set of writings to the other without distracting the reader's attention. Odum even timidly hinted that cooperative production methods and other forms of communal economic control might be necessary to overcome the ravages of underdevelopment. He did not, however, work the implications of such insights into his strategy of regional–national social planning. To surmount the misgivings apparently created by not facing this issue head-on and the problem of central State power, Odum and his colleagues often resorted to obfuscating literary ploys, but unlike the RPAA the southerners' analysis was not typically overshadowed by such rhetoric.

2. REGIONAL THEORY IN THE 1930s

The thing which set the work of the RPAA and Southern Regionalists apart from that of earlier planners was their focus on the *region* as a planning unit. This was not a matter of convenience or arbitrary choice, but a fundamental aspect of their world view and doctrine. For both the RPAA and the southerners, regions were conceived as the primary building blocks of human culture and social life. They were real historical places, such as New England, the South, or the Pacific Northwest, which shared a common history, social institutions, and patterns of human/environmental relationships. This is a markedly different concept from the regional definitions used by planners during the 1960s which have become enshrined in contemporary textbooks (see Chapter 5).

Mumford's descriptions of regions and regional life could well have been written by Vidal de la Blache or Demangeon and, indeed, were often used by later geographers as prime English-language examples of descriptive regional analysis (Dickinson, 1947, 1964; James and Jones, 1954). The region was described as a territorial community, not infrequently associated with long-standing political subdivisions or a city's extended hinterland, which had developed a unique sociographic identity. This unique character was the synthesis of particular attributes of the physical environment and the people who settled it, working themselves out in an intimate, symbiotic fashion. Borrowing Geddes' 'valley plan of civilization', there was a strong suggestion (e.g. Mumford, 1938) that occupations, settlement patterns, and land uses tended to sort themselves out within the region based on natural habitat characteristics and relative distance from the larger cities.

The model for such a world of human-scale regions was based on an interpretation of the social and economic geography of pre-industrial Europe. The RPAA tried to fit this model to regional life in America, alternatively analysing individual states like New York and Massachusetts, city regions such as Boston, and multistate groupings like New England (MacKaye, 1928; Mumford, 1938; Wright, 1926). The most important addition made by the RPAA to the European regional concept—largely attributable to MacKaye, but also Mumford—was the injection of wilderness areas as necessary elements of the regional mosaic. This was influenced by North American experience and the life's work and writings of John Muir (b.1838–d.1914) and George Perkins Marsh (1864).

The strong points of such a formulation were that it accented the relations between town and countryside, and focused on the contradictions in rural/urban relations created by capitalist urban-industrialization. Its central weakness was its failure to emphasize the class relations and class contradictions of capitalist industrialism, allowing the RPAA to slip rather precipitously into an overly organic view of the region and regional society. This metaphor led to quite literal comparisons between 'regional processes' and natural biological actions, and also tended to suggest physical barriers such as new towns and greenbelts to retard unwanted flows and movements. The idea of characteristic social relations associated with a particular mode of economic organization, complementing and contextuating geographic patterns of production and consumption, simply was not mentioned by MacKaye and received a meagre coverage of ten pages or so in Mumford's 500-page 1938 classic. This failure to acknowledge the class characteristics of American industrial expansion also led to an acceptance of government as a disinterested arbiter of regional problems.[5] Despite Mumford's recognition of the State's central place in international militarism, he and his fellow planners were willing to assign it a pivotal role in achieving 'regional balance' through a reconstructed industrial order.

The Southern Regionalists defined regions as substantial areas of the national territory sharing a common history and political identity. While they recognized the physical and economic dimensions of regionalism their main concern centred on social structure, cultural heritage, and political history. This concept was obviously drawn from the experience of the American South and, as already mentioned, shared much in common with the themes of French regionalists. The espousal of political regionalism was probably the most controversial component of regional theory during the 1930s and lends the southerners' thinking a poignancy and marked relevance to contemporary regional issues. There can be no question that regional planning for Howard Odum and his colleagues was an attempt to present regional political demands in a form palatable to the national establishment and that the Roosevelt administration's adoption of the idea was in some measure a response to southern sentiments, heightened in the midst of economic crisis.

In some ways the southerners' idea of regionalization was more sophisticated than that of the RPAA. Their emphasis on shared history and regional political consciousness, while maintaining the geographic palpability of the regional concept, shifted regional studies from the sphere of human ecology to that of sociology. This removed some of the nature-determinism and transcendentalism from the regional idea, as well as relieving somewhat the unending haggling over identification of regional boundaries. By readily acknowledging the social-psychological element in regionalization the Southern Regionalists set the stage for recognition of the crucial importance of political awareness and political organization for any successful strategy of regional planning. As I shall argue shortly, however, historical circumstances prevented southern planners from pursuing effective action along these lines.

The importance given to class and racial issues was another noteworthy feature of southern regionalism which set it apart from formulations of the RPAA. In the post-bellum South, only three generations after the Civil War, it was impossible to overlook the grinding toll taken by racism and class exploitation. Systematic ostracism and criminality towards 'free' blacks had been institutionalized into the southern legal system for several decades and had, in fact, been the first research interest of Odum's professional career. In cautious language, reflecting the reign of terror likely to be inflicted on overzealous southern liberals by their white compatriots, regional planners recognized that putting an end to southern apartheid practices would have to be one of the first steps towards any programme of regional reconstruction. This was a muted note within the planners' repertoire, though, and Odum hoped that it might be accomplished through exogenous, 'technical' reforms such as the construction projects initiated by the New Deal (Odum, 1939).

The plight of southern sharecroppers—perhaps a majority of the southern population of the period—can legitimately be said to have been the focus of most of the Southern Regionalists' strategies for achieving regional balance. The brutal system of forcing the returns of agricultural labour down to and below the subsistence level, exporting the surplus value realized through middleman profits northwards, was a primary obstacle to indigenous capital accumulation and an inward-looking expansion of the southern industrial structure. Little progress could be made towards improving southern living standards as long as the social relations of production vetoed any benefits which might accrue to rural working-class initiative. In addition, southern agricultural productivity was steadily falling behind the rest of the country because of disincentives imposed by the social system to adoption of less labour-intensive production methods. Regional planners saw the need to revise land tenure statutes and pursue cooperative methods for increasing social access to modern capital equipment, but given the generally reactionary spirit which pervaded most classes of southern society such communalist measures had to take a back seat to more technically oriented proposals dealing with agricultural research, extension, and adult education.

The major weaknesses of Southern Regionalist formulations, in comparison to those of the RPAA, were the fuzziness of their view of rural/urban contradictions and their tendency to construct a conspiracy theory of urban-industrialization. The RPAA, in their efforts to understand the mechanics of changing rural/urban relations, had traced in some detail the transformation of contradictions between town and countryside under industrial capitalism. They realized that the growth of large metropolitan centres and the bankruptcy of rural life were both part and parcel of commodity relations triumphant and the emerging ideology of a consumer society (although they offered few proposals which struck at the heart of the process). The southerners, on the other hand, understood the exploitation of southern surplus labour by northern-owned enclaves of secondary industry and were keenly aware of the connection between reserve labour and the workings of the rural sharecropping system. They tended to see urban-industrialization in terms of a new conflict between North and South, failing to perceive the underlying economic mechanics of the transition to industrial production. Odum (1934) responded to the apparent threat to rural-oriented southern life and cultural values by calling for a 'ruralization' of the city—as opposed to continued urbanization of the country-side—a move which in itself stood little chance of changing the nature of the urbanization process then underway.

3. PLANNING STRATEGIES

Planning strategies put forward by the Southern Regionalists and the RPAA emphasized mixing town and countryside through the decentralization of industry. This was to be based on an application of new technologies (especially energy and transport technologies), new town construction, and the reform of regional education and politics. The centrepiece of this strategy was the construction of new towns. This was particularly characteristic of RPAA proposals, figuring in almost all of their writings. The Southern Regionalists also endorsed the new town idea, but only when it had appeared in New Deal programmes such as the Tennessee Valley Authority and the Resettlement Administration.

In elaborating the new town concept Mumford (1938) paid tribute to the early formulations of Owen and Fourier and explicitly recognized that the interests of finance capital, dominating the national market, were in direct opposition to Geddes' ideal of a 'biotechnic' economy, which was to serve as the basis for new town life. The theoretical foundation of the RPAA's new town proposals rested on Howard's garden city (Bing, 1925; Mumford, 1925b, 1938; Purdom, 1925; Wright, 1925). This is to say, new towns were to be small, decentralized settlement units, enjoying communal land ownership to ensure proper physical planning and amortization of construction costs. Land, however, was not the productive basis of the new town economy. The 'new towns' actually designed by the RPAA members Henry Wright and Clarence Stein, such as Sunnyside Gardens, New York, and Radburn, New Jersey, were, in fact, 'garden suburbs', with no independent economic base (Stein, 1957). As with the grander-scale new town programme in Britain, initiated by the New Towns Act of 1946, much of the RPAA's discussion revolved around the design and quality of housing units. Mumford (1938) did conceive the planning of new town industrialization as part of a broader, socialized regional planning effort, but the real emphasis of the RPAA's new town arguments was in demonstrating their technological feasibility.

The Association's thinking on the economy of new towns centred on two specific technological innovations: large-scale electric power generation and modern highway construction (Bruere, 1925; Chase, 1925, 1933; MacKaye, 1931; Mumford, 1938). As was pointed out in Chapter 2, the predominant reasons for early industrial localization and concentration were access to resources, power, and markets. Location of industrial activities during the formative stages of the industrial revolution had been severely constrained to the immediate vicinity of resource inputs and trans-shipment points, setting the pattern for later urban-industrial development and regional specialization. Both the RPAA and the Southern Regionalists thought they saw a technical remedy for this historical impasse.

Introduction of the factory system of production had concentrated industrial activities in a revolutionary manner and allowed the development of major scale economies, based on the specialization and close supervision of labour and the invention of new machine technics. The new technology, however, required plentiful supplies of non-human power: water power, steam, and coal for boilers. These production requirements had exercised an incalculable influence on the geography of economic activities. Since the writings of Peter Kropotkin and Thorstein Veblen at the turn of the twentieth century, however, social critics had argued that new technological developments in power generation should allow this pattern to be modified.[6] Electrical energy that could be produced in quantity and transported at low costs would make universal accessibility to almost unlimited cheap power a real possibility. Rural electrification would not only allow factories to be moved out into the countryside but production itself could once again be dispersed away from the factory site, with its waning monopoly on the power necessary to run industrial machinery. At the limit, it might be conceivable that work activities could be focused once more on the human-scale workshop, and even be reintegrated into domestic life.

Advent of the motor car and the beginning of widespread highway construction complemented the revolutionary possibilities of rural electrification. Economic movements of goods, resources, and people which had been chained to water routes and then railway lines could at last be liberated from the tyranny of place. The idiosyncratic location of natural resources, ports, railroad terminals, and population centres would no longer dictate where secondary industrial processes could take place. This was a radical transformation, which, ideally, opened up every crossroad and village to active participation in the industrial economy.

These changes undeniably pushed back the potential boundaries of the industrial world and regional planners greeted the new era with a proclamation of broadening horizons and the advent of 'footloose' industry. As always, though, technology was a double-edged sword. Capable of cutting in either direction, its outcomes were highly affected by the nature of economic relationships and the whole structure of supporting social institutions. Planners recognized this in part, and regional planning itself was put forward as an institutional process which could guide the application of new technologies. However, while Kropotkin had called for fundamental changes in the purpose and social relations of production as a concommitant to technological change, planners rested their hopes more squarely on the transformational possibilities of technology itself. As with Odum's 'technical' solution to southern racism, they apparently hoped to by-pass the explosive political question of who controlled the economy. This was also evident in the political and education strategies adopted by the planners.

Both the Southern Regionalists and the RPAA had explicit commitments to new forms of political organization and the use of public education to obtain their goal of regional balance. While representing important advances over contemporary practice, both their political and educational views, however, skimmed lightly over the fundamental issues of real political power and its relationship to economic control.

In the educational sphere the RPAA took their cues from John Dewey and Patrick Geddes.[7] Both men had argued that education should be founded in immediate, real-life experience, and that such training formed the basis for later participation and rational choice in public life. This doctrine had a significant impact on both educational philosophy and approaches to liberal reform in the United States. Dewey felt that educated citizens, familiar with an empirical, learning response to their environment, would provide the impetus necessary to bring about major changes in civic consciousness and public decision-making processes. Geddes believed that a profound personal knowledge of the local environment, instilling a kind of regional consciousness, would not only provide citizens with the ideas and information necessary for their own life experiences but also serve as an irresistible prod to adoption of sound ecological principles and a planning approach to urban and regional renewal.[8] Thus Dewey and Geddes used an 'active' concept of education to bridge the gap between education and political life, and this was exactly the concept of education *cum* politics adopted by the RPAA. Mumford's 1938 discussion of 'The Politics of Regional Development' (*The Culture of Cities*, Chap. 7) first argued for a political framework of federated regions and then presented the 'Survey and Plan as Communal Education', a process that would 'progressively liquidate' the whole metropolitan regime.

Once again, the Southern Regionalists had more pedestrian notions of political and educational reform. The Progressive Movement, which had been fueled in part by Dewey's pragmatism, largely by-passed the South. The region's pre-industrial economy and social institutions were not receptive to such reforms. Odum and the other members of the New South Movement wanted to incorporate earlier Progressive reforms as part of their own programme of regional reconstruction. Political activism thus became defined as professionalizing State and local administration and adopting the other forms and trappings of modern representative government. In the southern context, of course, this could only be seen as genuine progress. Regional consciousness, meaning awareness of the South's dependent situation and systematic exploitation, was also a factor in the southerners' political strategy. Rather than being used as a rallying point of renewed regional militancy, however, it was meant as a subdued motive for supporting regional planning. Given northern feelings about southern 'sectionalism' this was all that could possibly find support from national public opinion, and the Southern Regionalists were meticulous to disassociate themselves from any sectionalist taint.

So while southern interest in education laid stress on regional history and rural values its consciousness-raising function stopped short of serious militant agitation (see Odum, 1934; Odum and Moore, 1938). Attention was focused on other important concerns: improving the basic quality of public instruction, providing programmes of extension and adult education, and significantly upgrading regional institutions of higher learning. These strategies were aimed towards radically changing general educational attainments—supposedly bringing with it an equally improved outlook—and gearing up the colleges and universities to (1) direct their R&D efforts (virtually non-existent at the time) towards isolating and solving regional problems and (2) provide appropriately trained cadres to manage the newly modernized system and social institutions. The suggestions offered by Odum, taken in this context, would be accounted sound in most parts of the world today and went far towards making Odum's home institution, the University of North Carolina, one of the first modern southern universities to attain a reputation of national stature. Not surprisingly, its planning school, founded by Odum at the end of his life, was long considered one of the world leaders.

Regional theory and strategies circa 1933, while not achieving the sophistication represented by Mumford's and Odum's 1938 publications, set out a fairly comprehensive programme. It condemned capitalist urban-industrialization as a cancer which was destroying regional life, ecological integrity, and even Western civilization. It proposed a planned approach to regional reconstruction, based on attaining a new equilibrium or regional balance, impacting both internal regional relationships and interregional linkages. The mechanisms for reaching these goals would be industrial decentralization, new town building, power and highway construction, and educational and political reform. The questions that planning doctrine did not squarely address were the bottom-line issues of economic control and political power. Planning experiences during the New Deal demonstrated the strengths and weaknesses of this platform.

4. REGIONAL PLANNING AND THE GREAT DEPRESSION

One of Franklin Roosevelt's first moves upon assuming the American presidency was to step up the flow of public works projects through the bureaucracy. These construction jobs were meant to provide stopgap employment for some of the millions of people out of work in 1933, and by the fourth year of the Depression the situation was becoming desperate. The National Planning Board (NPB) created to coordinate this operation soon defined a greatly expanded role for itself, based in large part on the regional planning doctrine analysed in the preceding section (US National Planning Board, 1934; US National Resources Committee, 1935, 1936–1943). Before the NPB's successor, the National Resources Planning Board (NRPB), was dismantled in 1943 much of the United States had been touched by some kind of regional planning activity. This ranged from organizing regional commissions and preparing resource surveys to dam building and comprehensive regional-resource development schemes. Most important for the present discussion were the greenbelt new towns and the river basin development projects, especially the Tennessee Valley Authority (TVA).[9]

The greenbelt towns, the most visible artifacts of the Resettlement Administration's work (Conkin, 1959; Friedmann and Weaver, 1979; Myhra, 1974), were planned largely by four RPAA members, including Henry Wright and Clarence Stein, and were singled out by Mumford (1938, pp. 392–401) as examples of the move towards garden city development. In the same discussion Mumford went on to decry the lack of any such greenbelt town construction in the Tennessee Valley area. Analysis of these projects provides a vantage point for examining the relationship between regional planning theory and practice during the 1930s. While regional theory suggested that new towns and resource development could help decentralize the metropolitan economy and recreate balanced regional environments, planning practice contributed to just the opposite result, increasing the degree of metropolitan concentration and control.

In April 1935 the Suburban Resettlement Bureau which had been struggling along for a year as an independent agency was consolidated with several other land-use planning endeavours to become the Resettlement Administration (RA). The concept was Rexford G. Tugwell's, the second member of the 'brain trust' which had helped Roosevelt since the beginning of his campaign for the presidential nomination. Tugwell intended the greenbelt new towns programme to provide housing and jobs for working-class people who were suffering the full burden of the economic crisis.[10] Before the RA was pronounced unconstitutional eighteen months later by the US District Court of Appeals in New Jersey three greenbelt towns were in varying stages of construction: Greenbelt in Maryland, Greenhills in Ohio, and Greendale in Wisconsin.

The greenbelt programme (along with the TVA) received strong approval from members of the RPAA, despite their otherwise lukewarm support for Roosevelt and the New Deal.[11] The Association had, in fact, disbanded as leading spokesmen drifted off to work for different government agencies and programmes. Such practical applications were had, however, only at the cost of the RPAA's *most central* theoretical ideals – i.e. decentralization and regional balance. The RA programme is a prime example.

Rexford Tugwell, the RA's sole administrator, hardly qualified as a pragmatist. He was committed to an interventionist economic programme resting on the analyses of Thorstein Veblen, and was used as the central target of reactionary New Deal critics and red-baiters until he left Washington in 1937 (Sternsher, 1964). The image followed Tugwell throughout his public life – as New York City's first planning director, governor of Puerto Rico, and first head of the University of Chicago's Program of Education and Research in Planning. As administrator of the RA, Tugwell gave jobs to RPAA members like Wright and Stein because of their technical experience as new town planners. He felt the RPAA's concept of regional planning was hopelessly romantic, however, and held its main propagandist, Lewis Mumford, in near contempt.[12] For Tugwell, the purpose of the greenbelt towns was 'to put houses and land and people together in such a way that props under our economic and social structure will be permanently strengthened' (Tugwell, 1936, p. 28). While he believed in the necessity of helping submarginal farmers in rural areas, even going so far as to adopt communalist production methods in some RA experiments in the South, Tugwell did not believe there was any way to get substantial numbers of houses and jobs together *outside of existing metropolitan areas*. Tugwell understood that without political control of the means of production—for him, meaning national planning of industry—even large-scale suburban housing projects were doubtful enterprises. The Resettlement Administration's numerous troubles and eventual demise proved his point.

The RA was not trying to reconstruct regional life; its goals, in fact, were just the reverse. The RA was meant to expand the metropolis: 'The conception of suburban resettlement came less from the garden city of England than from studies of our own population movements which showed steady growth in the periphery of the cities. . . . In other words, (Greenbelt) accepted a trend instead of trying to reverse it' (Tugwell, 1937, p. 43).

Even this proved too difficult a task. The regional planners of the RPAA found themselves working for an extension of metropolitan concentration and influence, which, although 'accepting a trend instead of trying to reverse it', proved infeasible without a clear political mandate to control economic activities and the location of industry. The Tennessee Valley Authority provides an even more dramatic example, for here a regionally organized branch of the central government was given actual control over some aspects of a regional economy.

In 1935 the National Resources Committee released a pathbreaking report entitled *The Regional Factors in National Planning*. This study, prepared under the supervision of John M. Gaus from the University of Wisconsin, accepted many of the tenets of regional planning theory as proposed by the RPAA and Southern Regionalists. Howard Odum's influence on the NRC's Technical Committee on Regional Planning which produced the publication was particularly important. The regional concept was developed as thoroughly, perhaps, as has ever been done in an official government document, and the idea of balanced regional resource development within the context of historically defined cultural areas was adopted as the explicit goal of planning. In attempting to provide a set of working criteria for the definition of planning regions, however, the Technical Committee fell back on the physical watershed or river basin as a reasonable approximation of a regional unit. This was largely a practical concession to the territorial division of powers within the American federal system and recognition of a number of ongoing river basin development schemes already initiated by the Roosevelt administration. It fits in neatly with the Geddes/Mumford 'valley section' idea, however, and made the apparent links between planning theory and practice appear stronger, perhaps, than they really were.

The bulk of *The Regional Factors in National Planning* was devoted to a discussion of four major river basin development projects and their associated regional planning organizations: the Connecticut River watershed and the New England Regional Commission, the Colorado River Basin Compact, the Pacific Northwest Regional Planning Commission and the *Columbia Basin Study*, and, finally, the Tennessee Valley Authority. The Colorado and Pacific Northwest efforts led to construction of the Boulder, Bonneville, and Grand

Coulee dams. These were all federal water control and power generation projects with, in truth, few broader regional planning pretensions. The TVA was, however, another story.

The Tennessee Valley experience had started with construction of a large dam and two munitions plants on the Tennessee River at Muscle Shoals, Alabama, during the First World War. After the War these facilities proved to be a political 'hot potato' because of the ideological implications of government ownership. The dispute remained unsolved until Franklin Roosevelt took the matter in hand, and at the beginning of the New Deal at least his latitude for manoeuvering seemed significantly broader than earlier planning advocates had even hoped for. Roosevelt proclaimed that 'this power development of war days leads logically to national planning for a complete river watershed involving many States and the future lives and welfare of millions' (Roosevelt, 1938, p. 122). The Tennessee Valley Authority Act of May 1933 gave the president power to make 'general plans' for the entire river basin and recommend legislation 'to realize a wide range of public purposes in the valley, including flood control, navigation, generation of electric power, proper use of marginal lands, reforestation, and "the economic and social well-being of the people"' (Derthick, 1974, p. 20; quoted in Friedmann and Weaver, 1979, p. 75).

Here, then, was an experiment which seemed to match the concepts and expectations of the theorists. It was acclaimed, quite literally, by both the RPAA and Southern Regionalists as the first steps towards a new civilization (e.g. see Chase, 1933; Mumford, 1938; Odum and Moore, 1938; Woofter, 1934). It was no accident that this project was undertaken in the heart of the American South. Besides the passive factor of existing publically owned infrastructure and the more generalized impetus of depression conditions prevailing throughout the country, regional consciousness and a growing commitment to regional planning among reform-minded regional elites had laid the groundwork for some such development initiative. This juxtaposition of regional demands, an explicit doctrine for regional planning, and a government caught in the throes of economic crisis allowed, perhaps, a temporary suspension of the prevailing norms of capitalist urban-industrialization.

Hints of any fundamental departure from the assumptions of liberal political economy—the appearance of a cooperative, socially planned regional society—were very short lived, however. David Lilienthal's (1944) 'dreamers with shovels' soon proved to be building yet another enclave of metropolitan America. How could it have been otherwise? The Washington government's problem was to revitalize the *national* economy, while maintaining civil obedience and territorial sovereignty. Any substantial realization of regional planning goals, despite theorists' claims to the contrary, would have significantly altered economic relationships and the balance of power in America. Even though some analysts have argued that TVA was coopted by traditional local interests (Selznick, 1949), there seems little doubt that evolution of the Valley Authority's role coincided with the interests of the Washington establishment.

In the early years of TVA there was scattered evidence that real change might be in order. Going beyond the timid proclamations of the regional theorists, planners in the Forestry Division and Regional Planning Division of the Authority's staff at Norris, Tennessee, seemed inclined towards a broad interpretation of the TVA's role. Foresters saw the problem of denuded land intimately tied to the southern land-tenure system and suggested programmes with a distinctly collectivist tinge. Regional planning, under the wing of the TVA Board's first chairman, Arthur Morgan, was conceived as an all-encompassing social task, but the Board's other two members never allowed the Regional Planning Division to set about its job.

There can be no question that, in part, these false starts were a matter of sorting out the ill-defined limits of TVA's legislative mandate. This has often been portrayed as a personal power struggle between Morgan and his successor, David Lilienthal, the coordinator of the TVA's power operation, responding to the changing political cues from Washington. More to the point, perhaps, it might be argued that cooperative ownership of forest resources and regionally based social planning would have been truly radical moves. They might have even provided a mechanism for regional control and planning of the Valley's water and electrical power resources, but this was never among Washington's intentions. Referring back to Roosevelt's speech on TVA potentials, quoted earlier, he called for 'national planning for a complete river watershed', not regional control of regional resources. And this is exactly what happened.

By leaving agricultural activities in the hands of traditional land grant colleges and state agricultural extension offices—the strategy criticized by Selznick in TVA and the Grass Roots—Washington left basic southern economic relations intact. The sharecropping system was propped up, and the supply of cheap cotton, surplus labour, and middleman profits was guaranteed by the dependent southern ruling class. All that was needed was to secure public control of electric power production, attracting northern industry to take advantage of cheap power and the region's unorganized, cut-rate labour.

While Progressives from the time of Senator George Norris had campaigned for public control of the Muscle Shoals electrical plant, their populist intentions had been to raise rural living standards by breaking the private utility monopolies and distributing cheap electricity. This meshed perfectly with regional planners' arguments about the use of electrical power in decentralization and achieving regional balance. Neither Norris nor the regional theorists, however, took into account the impact of class structure and the role of central government in such a situation.

In 1936, in a rare show of support for the Roosevelt administration's programmes, the US Supreme Court declared that it was legal for TVA to produce and sell power in competition with local utility companies. This proved the beginning of a chain of events that was to set the direction for all further developments in the Tennessee Valley. Rather than being an attack on the interests of big capital, public power production by the TVA represented the most direct possible support for large corporate interests. Existing private utility companies realized usurious but limited aggregate profits by burdening a small number of domestic consumers with high rates. The Supreme Court decision allowed Washington to compete with these local monopolists, guaranteeing an immediate demand for publically produced power. By undercutting private utilities TVA was able to attract customers and begin amortizing its investment, thus subsidizing the cost of eventually supplying energy to the big industrial producers it hoped to attract into the region. TVA consistently charged lower rates to large-scale industrial users than small customers (Shapley, 1976), perverting the supposed technological solution to urban-industrialization envisioned by regional theorists.

When Arthur Morgan's term at the helm of TVA expired in 1938 Roosevelt replaced him with David Lilienthal. From then on TVA became the 'Tennessee Valley Power Production and Flood Control Corporation' (Tugwell and Banfield, 1950, p. 50). The regional planning unit was eliminated altogether. The troublesome Forestry Division was put in the safe hands of the 'locally' controlled Agriculture Division and TVA confined itself to the business of stopping floods and producing cheap industrial power. The former helped stabilize sharecropping profits and the latter helped attract secondary export industries, both reinforcing the region's integration into the national economy.

Other empirical evidence gives strong support to such an interpretation. Taking what was to become an entirely different perspective on the regional development problem (see Chapter 5), Friedmann (1955) found that *the spatial structure of economic development* in the Tennessee Valley was essentially a pattern of urban-industrialization. By 1950, 20 per cent. of the region's workers were employed in manufacturing–an increase of over 80 per cent. above pre-depression levels (calculated from data in Friedmann, 1955, p. 21). One recent study attributed 'one-third of the new industrial jobs and at least one-half of the increased value added by industry' in the region to TVA investment (Robock, 1967, pp. 114–115; quoted in Friedmann and Weaver, 1979, p. 77). Similar trends could be observed in population growth as well, with almost all population increase in the TVA area between 1930 and 1950 going to metropolitan counties (Friedmann, 1955, pp. 55–57). The Tennessee Valley Authority itself was to move its base of operations to Knoxville when the opportunity arose, a move which has recently (1979) caused belated litigation in Washington.

River basin planning in the Tennessee Valley proved to be a potent instrument for reproducing existing economic relations and spreading urban-industrial development, but not for realizing the goals of regional planning theorists. Social and economic polarization increased, and vague notions of regional balance remained chimerical. What TVA did demonstrate was that the vital element left out of regional theory—community control of the economy—was not a tangential issue. It was crucial. Further proof of the TVA's essential failure to promote 'balanced regional development' was to be provided with the rebirth of American regional planning activities in the 1960s, when grinding poverty and black protest brought the Appalachian Regional Commission to basically the same area to deal with similar problems.

5. REGIONAL PLANNING AND THE WAR

The Second World War brought government interest in regional planning to an end in the United States. The immediate problems planning had been meant to solve were cured by the War itself; employment and national unity were effectively assured by military conscription, war production, and the confrontation with Japan and Germany. Various regional organizations continued on in one form or another, but most were dismantled during the war years and others foundered for lack of support in Washington. Some like the TVA managed to survive by drastically redefining their terms of reference.

Planning theorists, of course, continued to push their cause. The war was all-consuming, though. Most leading regionalists were pacifists and internationalists, but they were eventually drawn into supporting the war effort— mainly because of Nazi Germany's atrocities. By the late 1940s, however, planners had reemerged with now-familiar calls for decentralization, regionalism, and regional balance. Three noteworthy publications were to emerge from this renewed activity.

The first was a study prepared at Yale University (1947). Although it was directed by a French expatriot, Maurice Rotival, it was fundamentally a continuation of the work of the New England Regional Commission. The New England planning group had led an independent existence before being incorporated into the New Deal, beginning as the New England Regional Planning League in 1929, and this was yet another example of the New Englanders' commitment. The report was a combination of historically based regional consciousness and the planning doctrines of the RPAA.

A second book, this time an independent theoretical statement, was written by Ludwig Hilberseimer in 1945 and published in Chicago four years later. Hilberseimer, a German émigré architect of Bauhaus fame, adopted a Kropotkinesque title for his work, *The New Regional Pattern: Industries and Gardens, Workshops and Farms* (1949).[13] The contents followed closely in the same spirit. There was a definite central European quality about the book, a deeply rooted critique of capitalist development, especially finance capital, which took on almost Marxist tones. The first half of the book traced the course of city development from classical times and linked it with the transformations of economic production. Then the regional concept was discussed and the impact of capitalist industrialization on regional integrity was illustrated through arguments, photographs, maps, and paintings. Hilberseimer's major theme was the necessity to reintegrate town and countryside. Kropotkin, Geddes, Stuart Chase, and various American planning organizations were his major references.[14] The last section of Hilberseimer's work was a rather disappointing physical planning solution to the problems he had analysed. It was very involved, but showed little thought beyond the proposed design concept itself. One interesting note, however, was that the rationale for his carefully drawn regional settlement schemes was taken from Walter Christaller's *Die Zentralen Orte in Suddeutschland* (1933). This reliance on an abstract model of the regional 'space economy' foreshadowed a new paradigm of regional development and planning which I will discuss in Chapters 5 and 6.

The final publication of the era was also a mixture of things past and yet to come. As part of centennial activities in the State of Wisconsin a symposium was held at the University of Wisconsin in 1949 dealing with 'regionalism in America'. Funding was provided by the Rockefeller Foundation and a book of the same title detailing the symposium's proceedings was released in 1951 (Jensen, 1951). In some ways this was a last gathering of the faithful. Southern Regionalists Howard Odum and Rupert B. Vance were in attendance, as was John M. Gaus, who had directed the NRC's report on *The Regional Factors in National Planning* (US National Resources Committee, 1935).

The origins of regionalism were traced back historically to the eighteenth century, and due regard was paid to the French regionalists and the formulations of Paul Vidal de la Blache. American regionalism was examined in the South, Spanish Southwest, and the Pacific Northwest, and regional planning in the Tennessee Valley was described once again as an example of the promise of a regional perspective on national problems.[15] There was an undoubted link between the concerns which had prompted this conference and a long-standing interest by Wisconsin geographers and sociologists in rural development and central-place studies (e.g. Galpin, 1915). Like Hilberseimer's reference to Walter Christaller, however, this was a sign of things to come. The symposium's central focus—regionalism and its role in planning and administration—fell on deaf ears. National interests had been redirected towards the international scene and American regional planning in the traditional mould was a dead issue.

As we will see in the next two chapters, European needs for physical and economic reconstruction after the War gave a short-lived impetus to regional planning in Britain and France during the late 1940s and early '50s. Mark One new town construction began in the United Kingdom, and Gravier (1947) wrote his famous analysis of centralization and Parisian dominance in France. This was a minor aspect, however, of the hectic rebuilding activities of the post-war period, and it was not until the economic problems at the turn of the next decade that serious public interest was focused once again on regional issues in Western countries. By then new regional theories concerned with *promoting* urban-industrial development and economic growth had replaced the older planning ideas discussed in this and preceding chapters.

NOTES

1. This view conforms with the accounts given in many standards European sources such as Gravier (1947), Alden and Morgan (1974), and Hall (1975). Interviews with leading European authorities on regional policy and planning reinforced this

74

interpretation (see note 2 below). Despite the concerns which prompted the *Barlow Report* (1940) in the United Kingdom, no substantive action was taken on its recommendations until the late 1940s. In France and the rest of continental Europe regional planning was also a post-war endeavour. The Ruhr Regional Plan Authority in Germany during the 1920s and Dutch polder settlement during the 1930s are often cited as exceptions, but these activities were primarily large-scale land-use planning of the 'metropolitan' or 'local/regional' type.

2. What, if any, impact *The Culture of Cities* and other American publications had on the deliberations of the Royal Commission on the Distribution of the Industrial Population (The Barlow Commission, 1937–1940) in Great Britain is unclear. By the time the commission was convened there was already a widespread regional planning structure functioning in the United States, as well as several influential official publications, such as the US National Resources Committee's (1935) manifesto *The Regional Factors in National Planning*. Today there is fairly wide agreement among European specialists that Mumford's *The Culture of Cities* (1938) carried American ideas across the Atlantic (from interviews with Derek Diamond, London, 29 Nov. 1976; Torsten Hagerstrand, Lund, Sweden, 16 Nov. 1976; Peter Hall, Reading, United Kingdom, 22 Oct. 1976; Leo Klaassen, Rotterdam, 10 Nov. 1976; François Perroux, Los Angeles, 20–24 Nov. 1978; Walter Stöhr, London, 23 Oct. 1976). Peter Hall, during the interview just cited, argued that J.F. Gravier's French Classic *Paris et le Désert Français* (1947) was significantly influenced by Mumford. In the 1972 edition Mumford is cited twice, *and through Mumford* mention is made of Patrick Geddes (Gravier, 1972, pp. 21–79). Mumford, in 1961, called Gravier's work an 'admirable study of the problem of achieving urban and rural equilibrium: the fruit of two generations of scholarship in an area where the French had been pre-eminent' (Mumford, 1961, p. 599). Gravier's earliest French sources were the regional geographers discussed in Chapter 3, especially Demangeon. Ironically, Gravier never mentions either Proudhon or Charles-Brun. Either they got lost in the cross-cultural shuffle or Gravier chose to disassociate himself from their memory in France.

3. For a more detailed sketch of the RPAA as an organization and the ideas of its individual members, see Stein (1957, especially the introduction by Lewis Mumford), Sussman (1976, Chap. 1), and Friedmann and Weaver (1979, Chap. 2).

4. A striking example of what I am attempting to describe here can be obtained by making a comparative reading of Mumford (1961) *The City in History* and Childe (1936) *Man Makes Himself*—two books ostensibly covering much of the same material. Mumford's analysis of the rise of the city in Western antiquity is fuzzy and at times all but inpenetrable, while Childe's, losing little in richness and texture, is precise and even enthralling. Mumford is certainly eloquent and humane. The difference lies in analytic method and focus. White and White (1962, p. 236) characterized Mumford's approach as 'organic metaphysics'.

5. As an institutional economist and labour advocate, Stuart Chase was more aware of this dimension of the regional problematique, but he, too, was to endorse the mainstream position of the RPAA (see Chase, 1933, 1936).

6. The influence of Thorstein Veblen and his 'institutional' economic theories on the RPAA is traced in Friedmann and Weaver (1979, Chap. 2).

7. Dewey's contribution to early regional planners is also considered in more detail in Friedmann and Weaver (1979, Chapt. 2).

8. The similarity in doctrines was not entirely coincidental. Dewey and Geddes met and collaborated on several occasions, both in the United States and Scotland (see Boardman, 1978).

9. The steps of this evolutionary process as well as many of the other, related, New Deal planning initiatives are discussed in detail in Friedmann and Weaver (1979, Chap. 3). Note that the NPB went through four name changes during its decade-

long existence and that the National Planning Board (NPB), National Resources Board (NRB), National Resources Committee (NRC), and National Resources Planning Board (NRPB) all refer to essentially the same organization.

10. The Depression years were the first time in several decades that non-metropolitan population growth exceeded metropolitan increases in the United States, a situation which was not to be repeated until the 1970s (see Doherty, 1980; Fuguitt, Voss, and Doherty, 1979; and McCarthy and Morrison, 1979).

11. See Mumford (1938, p. 400), Mumford and Osborn (1972), and Stein's remarks in *Towards New Towns in America* (1957, p. 119). See also Graham, 1967.

12. Interview with Rexford Tugwell on 29 April 1977. While numerous analyses by and about Tugwell have appeared over the years, his essential contribution to social-science-based planning has yet to be recorded.

13. I wish to thank Henry Hightower, Ernst Gayden, and Barb Hendricks for helping me to learn about Ludwig Hilberseimer's work and background. An annotated bibliography of Hilberseimer's work is available in Spaeth (1981).

14. Curiously, Mumford's *The Culture of Cities* (1938) was only mentioned in the bibliography, almost an afterthought. This was strange, because *The New Regional Pattern* was a miniature of Mumford's earlier book, although Hilberseimer's writing was more concise and his arguments more cogent.

15. Eastern regionalists such as the RPAA and the New England Regional Commission were not involved in the Wisconsin symposium. Whatever the reasons for this, the meeting's emphasis on cultural and political regionalism was much more in line with Southern Regionalist traditions than the concerns of Mumford and his colleagues.

CHAPTER 5

Post-war Growth and the Reappearance of Regional Theory

Two of the most striking features of the early post-war period in both western Europe and North America were the strong feelings of national unity and the continuing economic expansion brought about by the War. Regional issues which had characterized the first half of the twentieth century and had been fanned into flame by the Great Depression seemed legacies of a bygone era. Probably the best-known planning effort of the late 1940s and '50s to draw upon the earlier generation of regional theories was the British Mark One new town programme. Despite its widespread international reputation, however, new town planning in the United Kingdom proved to be an anomaly. It was never able to define an explicit set of policy objectives for itself, nor gain a genuine commitment from either of the major political parties (Aldridge, 1979).

More in keeping with the spirit of the times was a new variant of regional theory. Rather than focusing on the contradictions between town and country-side and the negative side of urban-industrialization, the new regional theorists interpreted metropolitan expansion as the central mechanism of regional as well as national development. Most writers were entirely uncritical of the social and geographic contours of capitalist economic growth. If there was any problem it was learning how to speed the process along and spread it around more evenly. Public interest in even trying to *facilitate* the action of market forces at a subnational level had to wait until the decade of the 1960s, when it began to seem that some areas were particularly difficult to integrate into the mainstream of national economic life. Meanwhile, planning theorists turned much of their attention to the problems of urban and regional development in Third World countries, where there was more immediate interest in their ideas.

In this chapter I will first recount briefly the early achievements of the British new town programme and then present an analysis of the major components of regional theory and regional planning strategy of the 1960s. This will allow us

Portions of Chapter 5, Sections 2 and 3, have appeared in earlier form in *Geoforum*, Vol. 9, No. 6 (1978), pp. 397–413; and in Stöhr, W.B., and Taylor, D.R.F. (Eds) (1981), *Development from Above or Below*, pp. 73–105, John Wiley, Chichester, Sussex.

to see how very alien the two theories of regional planning really were, while also providing a perspective for comparing the *new town* and *growth centre* concepts which later came to be associated with one another. The apparent continuity was primarily illusory.

1. THE BRITISH NEW TOWNS[1]

Ebenezer Howard's idea of the garden city was discussed in some detail in Chapter 3. Despite the lack of substantial socioeconomic content Howard's thinking stood directly in the tradition of the utopian socialists, and garden cities were meant as a serious alternative to continued metropolitan growth and dominance. After publication of the first edition of Howard's book in 1898 the Garden City Association was formed three years later, and in 1903 the Pioneer Company created by Howard and his associates bought 4,000 acres of land in Hertfordshire to begin construction of Letchworth. By 1912 Frederic J. Osborn had joined the Letchworth staff and the garden city movement gained probably its most steadfast English propagandist. In 1917 Howard, Osborn, C.B. Purdom, and W.B. Taylor formed the National Garden Cities Committee, for whom Osborn wrote *New Towns after the War* (1918). Some writers have argued that when Howard single-handedly began work on his second garden city, Welwyn, in 1919, he effectively sabotaged the work of the National Garden Cities Committee by turning his own and his colleagues' attention from lobbying efforts back to the details of site planning (Aldridge, 1979, pp. 17–18).

Osborn's design for garden cities after the Great War was basically disregarded by the public authorities. Some 'garden suburbs' like Hampstead Garden Suburb and Wythenshawe were built between the wars, but the massive council housing estates criticized by Mumford were more typical. The London County Council's estates at Becontree and Dagenham eventually expanded to accommodate over 100,000 people between 1921 and 1932. By the time the Barlow Report appeared in 1940 recommending a regional approach to building suburban satellite towns the renewal of hostilities in Europe precluded any serious consideration of internal affairs. The Garden City and Town Planning Association did continue its work, however, changing its name in 1941 to the Town and Country Planning Association (TCPA) in an effort to broaden its appeal.

Abercrombie's famous *Greater London Plan, 1944* (1945) appeared the year the War ended, and designated ten specific sites for satellite towns within a twenty-five mile radius of the city centre, including Stevenage and Harlow which were later given the official nod as new town locations. The same year the new Labour government went on to set up the New Towns or Reith Committee, which besides Lord Reith included Osborn and several other TCPA members. Parliament was so anxious to enact a new towns bill that the Committee, to their disgruntlement, was only able to release its *Final Report* (July 1946) a week before passage of the New Town Act of 1946.

In November 1946 Stevenage, which had already been selected by Abercrombie, was designated as the first new town site, and between 1947 and 1950 thirteen more locations were selected. Table 5.1 lists all the Mark One new towns, giving their year of designation, location, existing population, and population growth target at designation. Nine of the fourteen Mark One towns were located in the South East and were meant to absorb overspill population from London. Four others served the same function for Newcastle and Glasgow. Total new town population was projected to increase from a pre-designation level of 130,000 to over 700,000.

Table 5.1   Mark One British new towns (from Aldridge, 1979, pp. 2 and 176–178)

New town	Designation year	Geographic location	Population at designation	Population target at designation
1. Aycliffe	1947	North East	60	10,000
2. Basildon	1949	South East	25,000	80,000 (1951)
3. Bracknell	1949	South East	5,000	25,000
4. Corby	1950	South East	15,700	40,000
5. Crawley	1947	South East	9,100	60,000
6. Cumbran	1949	Wales	12,000	55,000
7. Harlow	1947	South East	4,500	71, 000 (1961)
8. Hemel Hempstead	1947	South East	21,000	60,000
9. Peterlee	1948	North East	200	25,000
10. Stevenage	1946	South East	7,000	80,000 (1967)
11. Welwyn	1948	South East	18,500	36,500
12. Hatfield	1948	South East	8,500	25,000
13. East Kilbride	1947	Central Scotland	2,400	82,500
14. Glenrothes	1948	Central Scotland	1,100	55,000

After the initial burst of activity there was only one further designation, Cumbernauld in 1955, before the renewal of interest in new town building after passage of the second New Town Act in 1959.[2] The first designations of the Mark Two program, Skelmersdale and Livingston, came in 1961 and 1962. As we will see in Chapter 6, this decade-long lull in new town starts during the 1950s was accompanied by a similar retrenchment in most other regional planning measures as well.

The important thing to note here is that Mark One new towns had little significance in terms of national regional policy as envisioned in the Barlow Report. According to Aldridge (1979, pp. 176–177) only two of the first-generation towns, Peterlee and East Kilbride, had even a questionable intention of promoting 'regional regeneration'. Their primary concern was providing a decentralized, human-scale approach to meeting metropolitan housing needs. New town industrialization was encouraged in an attempt to meet Howard's garden city ideal of *self-containment*, but, in practice, had few broader regional connotations. By the end of the 1970s some forty new towns had been built, but they only accounted for less than 10 per cent. of all post-war development and redevelopment (Stretton, 1978, pp. 165–166).

2. NEW REGIONAL THEORIES: PLANNING AS SPATIAL DEVELOPMENT

If British new towning sought neither the pre-war American ideal of regional balance nor to promote economic growth in outlying regions, it was during the 1950s and early '60s that an entirely new focus for regional planning became established. In these years direct concern for economic development was combined with theories purporting to explain the location of economic activities, giving birth to regional science and spatial development planning. The explicit purpose of the new doctrine was to promote *regional economic growth* through induced urban-industrialization. On the American side, the main architects of this synthesis were Douglass North, Walter Isard, and John Friedmann (Friedmann, 1955; Isard, 1956; North, 1955). Working independently in Europe, François Perroux, Jacques Boudeville, and Jean Paelinck developed what was to become a complementary line of reasoning (Boudeville, 1960, 1961, 1966; Paelinck, 1965; Perroux, 1950, 1955).

Douglass North, an economic historian at the University of Washington, proposed that regional economic growth takes place in response to exogenous demand for regional resources. Although Alvin H. Hansen and Harvey S. Perloff had presented most of the fundamental components of North's argument over a decade earlier, their discussion of natural resources and regional development was not presented within the emerging economic development paradigm (see Hansen and Perloff, 1942).[3] North divided the regional economy into two sectors: basic and residentiary. The basic sector was said to be the propulsive force in economic growth, supplying needed inputs to the national economy and thus bringing outside wealth into the local economic system. Multiplier effects from the expansion of basic activities would lead to the development of local service industries, nourished by linkages to various stages of production for export and final demand for consumer goods.

The basic assumptions of North's model were twofold. First, growth is a function of external capital accumulation and subsequent interregional flows of revenue and capital and, second, trade in a market economy proceeds on the basis of comparative advantage and equal exchange. Although North's theory was stated in terms of trade relations between discrete regional units, the concepts of regional complementarity and increasing regional interdependence clearly pointed to a continuing process of functional integration of the national space economy. It was posited that the prosperity and well-being of people living in one region were, in fact, dependent upon the conscious elaboration of economic ties with people living in other areas.

Even given the conceptual simplicity of North's export-based notion, it has provided the touchstone for most subsequent theorizing about regional growth (Borts and Stein, 1962; Perloff *et al.*, 1960; Perrin, 1974; Richardson, 1969,

1973a, 1978a; Thompson, 1968). *Increasing the territorial division of labour, decreasing the friction of distance, and augmenting the level of interregional trade are still usually said to be the keys to local economic growth and development.* Stated another way, it is argued, or perhaps more correctly it is taken as a fundamental truth, that people cannot satisfy their own human needs and provide themselves with an increasing level of prosperity through their own labour and the use of their own resources. This is an extremely critical issue, based ultimately on a particular view of the creation of economic value, to which I will return in Chapter 8.

Walter Isard, the founder of the Regional Science Association, accomplished the feat of integrating the German tradition of location studies into the framework of neo-classical economics. Not withstanding prior efforts by Edgar Hoover (1937, 1948), it was not until the appearance of Isard's *Location and Space Economy* in 1956 that the English-speaking world began to consider seriously the spatial dimension of economic activities. Following the lead of Alfred Weber (1928), Isard placed the emphasis of his work on what he called the *transport inputs* to production. He argued that the cost of overcoming the friction of distance should be considered of equal importance with the traditionally recognized production factors: labour, resources, and capital. Applying the equilibrium doctrine of neo-classicism to the space economy, Isard argued that market mechanisms would arrange economic activities in their optimal, profit-maximizing locations, creating an hierarchical economic landscape based largely on substituting transport costs for other production inputs.

This was an abstract, open-system model of the space economy. Idiographic problems such as natural resource locations and political boundaries were largely assumed to be solved, leaving the two primal economic forces—production and consumption—to be balanced off in locational terms. All other things being equal, this logic suggested that, eventually, most economic activities should gravitate towards the same selected set of locations. Ultimately the locational problem would be solved through development of an urban network of *nodes* and *linkages*, closely resembling August Lösch's earlier formulation (1954).

Cities became point locations exerting varying amounts of economic attraction, depending primarily on their size. Their main locational characteristic was their relative situation *vis-à-vis* other places in a theoretically unlimited urban system. All connections with the concrete world of cities and regions with proper names and individual identities discussed here in earlier chapters were lost, etherealized into the n-dimensional realm of economic space.

Two critical assumptions lay at the heart of Isard's theory. First, to create such a neat functional ordering of economic activities it was necessary to postulate that all the factors of production are more or less footloose, moving around freely in response to private economic decisions. Second, these micro-scale decisions, meant to maximize profit at the level of the individual firm, must also contribute to achievement of aggregate-scale economic efficiency. It is clear that such an economy would be predominantly urban-centred. What is more, by definition, this would mean a high degree of polarization in the location of people, resources, and capital, as well as a marked dependence on relatively cheap transportation. Overcoming the friction of distance became the major imperative in an Isardian world.

John Friedmann was among the first to become interested in the policy implications of such a theory for regional economic development. If aggregate economic efficiency was dependent on a highly integrated urban economy, then the best way to achieve regional economic growth was through encouraging development of the urban system. This meant that *regional planning* must become *spatial systems planning* and that the main concern of planners should be optimizing the location of economic activities. In 1955 Friedmann wrote:

> Linkages among city regions extend into all directions, joining dominant city to dominant city, sub-centre to sub-centre. The economic relationships which are expressed by these linkages in an advanced stage of economic development make up the central subject matter of regional economic planning *beyond* the immediate boundaries of the city region itself.
>
> These interrelationships are characteristic of any economy which is organized on a modular pattern with functional differentiation among its parts. . . . This system is the structural framework within which economic development takes place and which, at any time, places the maximum limits on the extent of possible development in any particular segment of the system (Friedmann, 1955, p. 143).

Several years later Friedmann further developed the logic of this argument in what became a widely cited definition:

> . . . *regional planning is the process of formulating and clarifying social objectives in the ordering of activities in supra-urban space.* . . . This formulation links regional planning to its basis in the pure theory of location . . . (Friedmann, 1963, p. 64, emphasis in the original).

In emphasizing the importance of locational decision making to regional development Friedmann had cast his lot with the budding multidisciplinary field of regional science. In doing so he unavoidably borrowed its central intellectual traditions. Regional planning as a field of study and professional practice was to interest itself in economic location theory, central-place studies, urbanization, and regional economic development. Its methods were to be rigorously scientific and its goal was to be the functional integration of the space economy, concentrating people, resources, and economic activities into a tightly woven network of cities and their adjoining regions. As we will see below, this notion of spatial systems planning, when combined with corollary ideas about scale economies, polarization, and unequal development, has provided the dominant framework for debate about regional growth until the present day.

As should be apparent from the foregoing discussion, there was a basic internal contradiction among the several components of this new breed of regional theory. The line of reasoning began with North's proposition that economic growth takes place through the stimulus of trade between different regions. What North had in mind were discrete, bounded, territorial units which would set in motion a process of *spatial interaction*, based on complementary economic resources. Isard's theory of location and the space economy focused on the flows and linkages between these different places, recognizing the overwhelming importance of cities in the industrial economy but abstracting them out of recognizable geographic space into a mathematical world of nodes and networks. Friedmann grafted the North/Isard models together to forge an eclectic strategy for regional development, capturing the manifold benefits of trade and regional complementarity by expanding and tightening the bonds of the urban spatial system. In the several steps of this process, territorial regions were overlain with a functional network of structural economic relationships, while the original buildings blocks—the regions—faded from view. The problem, of course, is that with them went the initial logic of the whole edifice; the export-base concept of exogenously induced growth was replaced by an *integrated system* which feeds on its own dynamism. In retrospect, development either became a *zero-sum game*, with some places and therefore some people benefitting from other peoples' labour and resources, or it was, in fact, a *self-initiating process* which could just as well have taken place within the confines of the originally postulated regions. We will come back to these questions in Chapter 8.

At the time, neither of these alternative interpretations were given centre stage, but the subsequent history of regional studies demonstrates the discrepancy between the *territorial* and *functional* elements of the theory. In practice, analysts either had to choose to deal with one part or the other; they either had to concentrate on regional–territorial characteristics (Alonso, 1968; Borts and Stein, 1962; Hirschman, 1958; Klaassen, 1965; Perloff *et al.*, 1960; Siebert, 1969) or opt for the spatial–functional approach (Berry, 1967, 1971, 1973a; Duncan *et al.*, 1960; Friedmann, 1955; Hansen, 1970; Hermansen, 1971; Pedersen, 1974; Richardson, 1973a). During the decade of the 1960s the spatial–functional school came into increasing ascendence, but both sides tended to look upon their differences as basically methodological problems, problems which could be worked out as regional science techniques became more sophisticated. It was not until the fundamental ideological assumptions of polarized development came to be challenged during the late 1960s that the theoretical nature of the conflict began to surface.

3. POLARIZED DEVELOPMENT AND GROWTH POLE DOCTRINE

Regional development theory did not stop with the idea of integrating the space economy, and regional planning needed more specific objectives if it were to transcend the level of arcane philosophizing. It was from an interest in polarized development and growth pole theory that conceptual evolution continued. In 1957 Gunnar Myrdal wrote a modest-sized book entitled *Economic Theory and Underdeveloped Regions*. Myrdal's well-known theme was that economic development, having started in certain favoured locations, would continue through a process of *circular and cumulative causation*. From a geographic standpoint, growth would be transmitted through a network of *spread and backwash effects*, meaning, respectively, the positive and negative impacts of continued growth in the original areas on other regions. Myrdal displayed a strong concern that the cumulative advantages experienced in the initial growth areas would cause backwash effects to prevail in most other places and that conscious policy intervention would be necessary to prevent a truly explosive global political situation. This was one of the first troubled references to ethics and politics which entered post Second World War economic theorizing, but it was largely ignored.

In the following year, Albert Hirschman published an independently conceived essay covering many of the same ideas. Hirschman's views were more in line, though, with the equilibrium arguments which dominated the contemporary intellectual scene. In *The Strategy of Economic Development* (1958), he spoke of *trickle down* and *polarization* instead of spread and backwash, but the concepts were fundamentally the same as Myrdal's. Hirschman, however, had much greater faith in the efficiency of market forces for allocating the factors of production and believed that while development might tend to polarize around certain initial *growth centres*, eventually trickle-down effects would become predominant in the never-ending search for resources and new markets. This difference of interpretation between Myrdal and Hirschman became a classic point of contention among planners.

Hirschman's argument fit neatly into the emerging paradigm of regional studies. According to his own summary:

> . . . we may take it for granted that economic progress does not appear everywhere at the same time and that once it has appeared powerful forces make for a spatial concentration of economic growth around the initial starting points. Why substantial gains may be reaped from overcoming the 'friction of space' through agglomeration has been analyzed in detail by the economic theory of location. In addition to the locational advantages offered by *existing* settlements others come from nearness to a *growing* centre where an 'industrial atmosphere' has come into being with its special receptivity to innovations and enterprise. . . .
> . . . there can be little doubt that an economy, to lift itself to higher income levels, must and will first develop within itself one or several regional centres of economic strength. This need for the emergence of 'growing points' or 'growth poles' in the course of the development process means that international and inter-regional inequality of growth is an inevitable concomitant and condition of growth itself.
> Thus, in the geographic sense, growth is necessarily unbalanced. However, while the regional setting reveals unbalanced growth at its most obvious, it perhaps does not show it at its best. In analyzing the process of unbalanced growth, we could always show that an advance at one point sets up pressures, tensions, and compulsions towards growth at subsequent points (Hirschman, 1958, pp. 183–184).

These paragraphs set the tone for regional development and planning for more than a decade. They contain three important contentions. First, because of the friction of distance, agglomeration economies would cause growth to be polarized in certain existing locations at the outset. Second, although by its very nature growth must be unequal or unbalanced, it also contains within itself the imperatives for geographic expansion. And third, such expansion will take place through the emergence of subsequent growth points or growth poles. Because of their importance, I will briefly trace the evolution of each of these ideas as they pertain to regional planning.

To begin with the agglomeration economies concept, Hirschman equated it with overcoming the friction of distance—getting everything close together to create a common pool of skilled labour, services, information, and infrastructure. In later years, economists and geographers have subdivided external economies into a bewildering host of categories, but the fundamental idea is that it is cheaper to be together. Sometimes this principle has been set at odds with the idea of scale of economies internal to the firm, but more typically it is argued that large plants require large urban markets, meaning big cities and an integrated space economy. No one has ever been able to measure scale economies very precisely, especially as they apply to the external environment. However, this line of reasoning led directly to the consideration of city size as a measure of scale advantages, and eventually to theories about optimum city size and the optimum shape of city-size distributions (Alonso, 1971, 1972; Berry, 1961, 1971; El-Shakhs, 1965, 1972; Mera, 1973; Neutze, 1965; Richardson, 1972, 1973b; Thompson, 1965). These two sets of ideas are not necessarily mutually supportive, but they have managed to coexist very fruitfully, providing a partial rationale for growth pole doctrine.

Turning to growth poles, as Hirschman had been aware, this idea originated with the celebrated French economist François Perroux. Perroux's argument was that leading industrial sectors could act as strategic *pôles de croissance* within interindustrial economic space, starting a process of self-sustaining economic growth which would radiate throughout the economy. Perroux's colleague, Boudeville, helped to transform the growth pole notion into a concept applicable in geographic space. Hirschman had already made this connection, and with later contributions by Lloyd Rodwin (1963) and Friedmann (1966), among others, growth centres provided the necessary link between theories of unequal development and the idea of inducing regional growth through integration of the space economy. The eventual logic of growth pole doctrine ran something like the following: disparities in welfare between different regions may be overcome by extending the polarized development process into depressed areas, through establishment of growth centres which link such areas to the economic growth impulses generated within the broader urban system.

Without becoming mired in the subsequent debates among growth pole theorists (see Boisier, 1980; Darwent, 1969; Friedmann and Weaver, 1979; Hansen, 1967, 1972; Kuklinski, 1972; Kuklinski and Petrella, 1972; Lasuén, 1969; Mosely, 1974), the key questions here are: (1) What is the true nature of polarized development? and (2) Are growth pole policies really a means of transferring economic growth to new areas (spread effects) or are they, in fact, misdirected, reinforcing the cumulative effects of polarized development (backwash effects)? There are no definitive answers drawn from planning practice to these questions, but most analysts at first tended to side with Hirschman's more optimistic view. Two of the most important studies by Borts and Stein (1962) and Williamson (1965) presented strong positive arguments. Borts and Stein attempted to show that eventual equilibrium in a space economy was inevitable, because capital would necessarily move outwards from core areas with a high K/L ratio in search of higher marginal returns on investments. Conversely, labour would tend to migrate from low productivity areas in response to higher wage incentives. In the end everyone would get their fair share.

Williamson's argument was that interregional revenue disparities could be conceptualized on a space/time continuum in the form of a bell-shaped curve. Polarization during the initial stages of the growth process would cause increasing regional inequalities, but at some unspecified point continuing polarization would naturally push things over the top of the curve and regional inequalities would again decrease. The secular trend would eventually point to an equalization of regional incomes.

Interestingly, Perloff *et al.* (1960), in one of the most impressive attempts at empirical analysis, had taken a very different view of similar evidence. While this classic study stood outside the prevailing imagery of polarized development, its conclusions are very pertinent: 'Economic decline in an area is an extremely difficult thing to face, particularly in our growth-minded culture, and yet the relative decline of certain areas in the *volume* of economic activities is an inevitable feature of a rapidly changing economy' (Perloff *et al.*, 1960, p. 607). This appraisal seemed to suggest that national economic growth would lead to two separate processes at the sub-national or regional level. Some areas, perhaps changing over time in response to changes in the structure of the national economy, would grow and become increasingly wealthy, while other areas would experience an absolute decline in their volume of economic activities. This could throw an altogether different light on the apparent convergence in regional income noted by Williamson, and, indeed, historical information continues to be interpreted as supporting either line of argument.

After lengthy practical experience in Brazil, Venezuela, and Chile, Friedmann reemphasized the less-optimistic view, drawing on the same intuitions as several emerging Latin American writers. In his 'A general theory of polarized development' (1972, original 1967) Friedmann attempted to conceptualize all the various factors—cultural, political, and economic—which enter into regional growth, and tried to anticipate the results of a policy of induced urbanization. His position was equivocal but somewhat disturbing. Polarized development appeared as a predominantly political process, with a dominant core area systematically exploiting its surrounding periphery through a monopoly of information and political power. He argued that eventually a *crisis of transition* would occur, either leading to a diffusion of political power and economic opportunity *or* ending in continued exploitation and possible political revolution. It seemed that at last Myrdal's original fears were gaining wider currency: regional development might indeed be a fundamentally ethical/political process and the ideology of polarized development might be a dangerous ploy. This was only the beginning of a growing critique of unequal development which will be discussed in Chapter 7.[4] It is important to note here, however, that as soon as the tenets of polarized development theory came seriously to be called into question, many of the central concerns of the precursors of regional planning and the first generation of regional planners began to reappear, eventually leading once again to a head-on attack on the fundamental mechanisms of capitalist urban-industrialization.

4. STRATEGIES OF REGIONAL DEVELOPMENT

The spatial development theories of the 1950s and '60s prompted planners in industrialized countries like France, Britain, and the United States to conceive the regional problem in terms of *depressed* or *lagging* areas. These 'regions' were typically delimited fairly arbitrarily as administrative or data-collection districts which displayed some specified combination of slow economic growth, low incomes, and high unemployment. Simply stated, they were not sharing in the more general growth and affluence of the national economy. Two approaches were suggested for dealing with such problem areas. The first, typically focusing on the most evident 'backward' regions, adopted a 'worst first' welfare strategy of *area development*. The second attempted to pinpoint the more dynamic urban places in lagging regions and use these as *growth centres* which could establish stronger linkages with the national core area, capturing the spread effects of national economic expansion.

Probably the best-known model for identifying and dealing with backward areas was set out in a series of OECD publications by Leo Klaassen (1965, 1967, 1968) of the Netherlands Economic Institute. Klaassen proposed to treat regions as discrete microcosms—area economies which could be understood through a study of the same kinds of accounting measures invented for application to the national economy by Colin Clark (1938) and Simon Kuznets (1941). Using the most common of the economic accounting measures, income, Klaassen suggested a simple benchmark technique, comparing area income level and growth with that of the national economy, as shown in Table 5.2. Three kinds of problem areas were distinguished (categories II, III and IV in Table 5.2): type II areas, with relatively low income levels but good growth rates; type III areas, with high income but poor growth performance; and type IV areas, which suffered from both comparatively low income levels and growth rates. The latter areas were the ones of most concern to politicians and planners.

Given widespread acceptance of the regional convergence hypothesis discussed in the last section, type IV areas could only be treated as anomalies.

Table 5.2   Klaassen's model for identifying depressed regions (From Klaassen, 1965, p. 30; reproduced in Friedmann and Weaver, 1979, p. 141)

Rate of increase in income compared to the national rate of increase	Income level compared to the national level	
	High ($\geqslant 1$)	Low ($\leqslant 1$)
High ($\geqslant 1$)	I Prosperity area	II Distressed area in process of development
Low ($\leqslant 1$)	III Declining prosperity area (potential distress)	IV Distressed area

Because of an unfortunate juxtaposition of circumstances such places were excluded from real participation in the national economy. In terms of a rational economic calculus, because of their own internal failings, they were unable to compete with other areas in putting together a workable factor mix in at least some branch of industry. They had no attractive resources—none that had not already been exploited at any rate. The local labour force did not possess the skills required by modern industry, and local capital formation was all but non-existent. Thus productivity levels were dismally low. It would be a travesty against national economic efficiency to reroute substantial amounts of scarce capital to such areas, so all that could be done—if political considerations made it necessary—was to spend some limited amount of money from the public coffers to improve the basic level of local infrastructure and social services. There might perhaps be some productive investment, but the keynote was welfare payments and retraining.

The ultimate logic of such programmes was made clear by Klaassen and Drewe (1973). In areas that could not compete, if anything was to be done at all by government, the working age population should be retrained and moved out to other regions where they could make a productive contribution to the economy. (The old and the young could be put on the dole.) It would be less expensive to subsidize corporate labour costs in core areas through public-supported relocation programmes than to attempt to make direct subsidies to capital in unfavourable locations. People prosperity was the important issue, not the prosperity of places. The real problem was that individual people responded imperfectly to market signals, so perhaps government was justified in helping them to become properly footloose.

As we will see in the next chapter, France, Britain, and the United States all adopted some variant of the worse-first area development strategy, but in practice these were mixed rather ambiguously with spatial development policies based on growth centres ideas. In France growth centres were identified with *métropôles d'équilibre* (Hansen, 1968) and in the United States with Economic Development Administration growth centres programmes and the Action Plans of several of the state/federal Title V Commissions (Berry, 1969,

1972, 1973a; Hansen, 1967, 1970, 1972). In the United Kingdom growth centres strategy came mainly to be imbedded in the Mark Two new town programme (Aldridge, 1979; Hall *et al.*, 1973; Hall and Day, 1978).

The unfolding of these programmes and their evaluation by mainstream regional scientists will be discussed in Chapter 6. Here it is useful to demonstrate more concretely the significant changes implied in the new regional development strategies by showing the confusion created in the British new towns programme with introduction of the growth centres concept. In part, the contradictions apparent in new town objectives during the 1960s and '70s were a result of attempts to overlay two opposing theories of regional planning.

As we already saw, Mark One new towning was grounded conceptually in the garden city notion, and aimed towards intraregional decentralization and some measure of non-metropolitan self-containment. Implicit ties with first-generation regional theory were very strong, although the broader concept of regional balance espoused by American planners was not incorporated as part of post-war new town doctrine.[5] Mark Two new towns came to life in a planning environment marked by renewed concern for regional issues, but presenting an unreconciled quiltwork of area development, growth centres, and garden city objectives. The resulting confusion led to contradictory policies and not infrequently rumpled tempers (see Aldridge, 1979, p. 141).

Not only were cross-purposes evident in conflicting departmental programmes but at a theoretical level the mixed garden city and growth centre ideas assigned to new towns such as Telford were at least in partial contradiction with one another. Table 5.3, garden cities versus growth centres, outlines some of the basic problems. While the contrasts presented in Table 5.3 are undoubtedly overblown, it can be argued that they point towards the fundamental incompatibility of the planning concepts at the heart of the two historical models of regional development.

At the most general level garden cities and growth centres share the goal of decentralizing national settlement patterns, but on almost all other points the two concepts are in basic opposition to each other. Working towards one set of objectives will sidetrack—if not preclude—an attempt to meet the other. This becomes doubly true when public spending limitations for regional planning are kept in mind. In theory, garden cities were meant to meet new housing needs in a carefully planned human-scale environment. This meant designation of smaller urban centres, or entirely new locations, on the periphery of growing metropolitan regions. The focus was on collective consumption, especially single-family housing, and physical planning. To avoid contributing to further metropolitan sprawl, self-containment, limiting the necessary commuting and contact with the outside, was to be achieved by creating a local service economy which would provide for a large proportion of local employment and consumption needs. This would be a 'balanced' social environment, with a cross-section of the typical national class structure, requiring a planned in-migration of residents representing different income groups. The fiscal viability of garden cities as an alternative to continued metropolitan growth would be demon-

Table 5.3 Garden cities versus growth centres

Garden cities	Growth centres
Objectives	*Objectives*
Creation of a human-scale environment	Geographic concentration of public investment
Meeting housing needs	Stimulation of growth opportunities
Improving public health	Development of comparative advantage
Metropolitan decentralization	Propagation of areal and hierarchical 'spread effects'
Self-containment	Integration of the national space economy
'Balance'	Normalization of the national city-size distribution
Partial regional closure	Labour absorption
Balanced public budget	
Implementation strategies	*Implementation strategies*
Designation of smaller centres	Selection of medium-sized urban places
Location in growth regions	Location in depressed regions
Focus on collective consumption, especially single-family housing	Public investment in industrial infrastructure
Emphasis on physical planning	Attraction of private investment through public subsidies to capital
Creation of a viable 'service economy'	Development of interregional linkages
Establishment of a representative cross-section of social classes	Concentration on selected 'basic' export activities
Limiting necessary commuting and physical expansion	Employment of existing workforce
Fiscal conservatism (complete mid-term return on public investment)	Concentration on economic and physical expansion
	Public costs balanced against net economic growth (private profitability)

strated by making new towns pay for their own way.

Growth centres, on the other hand, were meant as a mechanism for stimulating economic growth in depressed regions by concentrating available public investment funds. The idea was to develop comparative advantage in a selected group of export activities which could benefit from local surplus labour and begin sending spread effects through the depressed regional economy. This would help integrate the area into the national space economy by developing interregional linkages with other areas, and eventually help to normalize the national city-size distribution. The focus of new town investment in the growth centres mode would be to spend public funds in medium-sized urban places, emphasizing public investment in industrial infrastructure and direct subsidies to private capital. Since net economic growth was the ultimate goal—eventually raising substantially increased taxes as well as boosting regional incomes—public costs would be balanced off against private profitability.

Mixing garden city and growth centre policies in the same regional programmes meant that evaluating their success would be particularly difficult;

not only were funds being divided between two different sets of objectives but whatever successes the opposing policies were having could well be 'pulling' in quite different directions. On the whole, however, most 1960s regional policies in France, Britain, and the United States fell under the influence of the new generation of regional theories. These programmes are described and analysed in the next chapter.

NOTES

1. Book-length accounts of the British new town experience are available in Purdom (1949), Rodwin (1956), Osborn and Whittick (1969), Aldridge (1979), and Cullingworth (1980). My analysis here is heavily influenced by Aldridge's recent critical interpretation. Mumford and Osborn's collected letters (Mumford and Osborn, 1971) also give many interesting insights into the nuances of TCPA affairs and garden city theorizing and lobbying.
2. Cumbernauld actually gained designation through the Clyde Valley Plan of 1949.
3. There were other forerunners to North's argument; see, for example: Mackintosh (1923), Innis (1930), Lower (1933), Ahlmann, Ekstedt, and Jonsson (1934), Daley (1940), Graham (1940), William-Olsson (1941), Machlup (1943), Vining (1946), Stolper (1947), Klaassen *et al.* (1949), Dusenberry (1950), Hildebrand and Mace (1950), US Federal Reserve Bank (1952), Viner (1952), Andrews (1953), Meier (1953), Alexander (1954), Hoyt (1954), Isard and Peck (1954), Leven (1954), and Roterus and Wesley (1955). While these studies dealt with the difference between basic and non-basic activities or the impact of trade on economic growth, they almost always applied the concepts to either the *urban* economy or *international* trade. North's original contribution was to explicitly use the economic base concept in a regional context and to attempt to apply it as a model for growth in the 'ideal' North American region.
4. For further discussion of recent contributions to mainstream regional theory, see Richardson (1973a, 1978b, 1979), Lasuén (1974), Friedmann (1975), Stohr and Todtling (1977), Rodwin (1978), Friedmann and Weaver (1979, pp. 89–159), and Hansen (1981).
5. Rather curiously, given Mumford's close adherence to Geddes' thinking, Peter Hall contends that the regional concept developed by American planners during the 1920s and '30s was incomprehensible in England (personal interview, University of Reading, 22 Oct. 1976), despite the fact that Mumford's (1938) *The Culture of Cities* became the 'bible' of a whole generation of regional planners (Hall, 1975, p. 66).

CHAPTER 6

The Rise and Fall of Regional Policy

Second-generation regional theory remained in limbo for several years. While the glow of post-war economic growth waxed and waned there seemed little desire to interfere in the process. In the United States attention was focused on the Cold War and demonstrating the merits of the free enterprise system, although there were several attempts to push area redevelopment legislation through the US Congress. The earliest of these dated back to the 1945 Hays–Bailey Bill, emphasizing depressed farming areas, and the Murray–Sparkman Bill of 1949. The best-known effort was associated with Senator Paul H. Douglas, starting in 1955. His first bill, S.2663, was given strong support by John F. Kennedy who served as the measure's floor manager (Schlesinger, 1965, p. 572), but it failed to get out of committee in House form.[1] Nothing more happened until the turn of the decade.

In Europe the strongest post-war interest in regional planning appeared in Britain. The new town programme has already been discussed in the last chapter. New towns were augmented by a much-celebrated round of parliamentary actions between 1945 and 1950 which adopted in one form or another many of the depression-spawned recommendations of the Barlow Report (*Barlow Report*, 1940), basically resurrecting the 1934 Special Areas (Development and Improvement) Act (Law, 1980, pp. 45–82). This was really a stillborn effort. After 1950 at the latest the threat of regionally concentrated unemployment of the pre-war variety lost its immediate credibility, as even declining traditional industries experienced a brief stay of execution based on wartime destruction and pent-up consumer demand (McCallum, 1979, pp. 9–10; Law, 1980, p. 48). During the 1950s a meagre four to five million pounds per year were spent for regional development, and industrial decentralization was, effectively, put in abeyance (Glasson, 1974, pp. 178–184).

1. THE REBIRTH OF REGIONAL PLANNING

By the late 1950s post-war optimism about measures like monetary and fiscal policy, national accounting, and other macro-economic techniques as guides to steady industrial expansion had come into question. The sluggish international economy experienced a continuing recession at the turn of the decade, and John Kennedy, as the new American President, felt that an important aspect of the problem was geographically concentrated unemployment, caused by relocation of important industries which had been traditionally identified with specific regions. Kennedy's concern was partially fired by the acute problems experienced by his native New England, but he also had a more theoretical justification for his approach, founded on a belief in the fundamental significance of structural considerations to successful economic policy. This perspective had a long heritage among liberal Democrats, going back to its institutionalist roots during the first New Deal discussed in Chapter 4 (Schlesinger, 1965, p. 575).

In 1961 the Area Redevelopment Act was passed by Congress, establishing the first official American interest in explicitly sub-national economic policy since the National Resources Planning Board was dismantled by a frightened Congress in 1943 and the Tennessee Valley Authority was gradually transformed by David Lilienthal into the 'Tennessee Valley Power Production and Flood Control Corporation'. JFK's assassination in 1963 brought a forced transition from the New Frontier to Lyndon Johnson's Great Society, but Johnson's unique legislative prowess and his early faithfulness to his predecessor's policies continued the trend towards a rebirth of regional planning.

In 1965, before the obsession with Vietnam had quashed serious attention to domestic affairs, Johnson managed to push several regional measures through Congress, including the Appalachian Regional Development Act and its required sweetener, the Public Works and Economic Development Act of 1965 (PWEDA) (Derthick, 1974, p. 108). Once again America's second major historic cultural area, the Old South, acted as a strong impetus for a regional perspective on national economic problems, just as it had thirty years earlier under Franklin Roosevelt's administration. By the end of the decade a whole series of economic development districts and regional commissions blanketted the country (Cameron, 1970; Cumberland, 1971). While this was partly a token measure to prevent cries of sectional favouritism over the Appalachian Regional Commission (ARC) (Derthick, 1974, p. 108), along with the newly discovered crisis in the big cities it helped give the War on Poverty a distinctly regional and local focus in terms of its targets for social and economic policy and planning.

A somewhat different but analogous pattern of events was taking place in Britain and France. In Britain, the same economic symptoms which had set Kennedy in search of answers to regional unemployment now encouraged Labour and Conservative governments alike to rearrange their priorities. Responding to the 1957 recession, the 1958 Distribution of Industry (Industrial Finance) Act was soon followed up by the Tory's 'worst first' district policy (Local Employment Act of 1960) and culminated in government White Papers on Central Scotland and the English North East (1963). The Labour victory in 1964 brought an even more serious commitment to the problem and a drastic change in emphasis, with a whole series of new measures being enacted between 1965 and 1969 which gave real teeth to the industrial location and the area assistance machinery (Dunford, Geddes, and Perrons, 1980; Hall, 1975, pp. 140–151; McCallum, 1979). By 1968–1969 assistance to officially recognized Development Areas had reached 300 million pounds, a tenfold increase over the year that Labour came to power (Bray, 1970, p. 148), and the Mark Two new town programme was well underway (Aldridge, 1979; Cullingworth, 1980; Rodwin, 1970).

Across the Channel events followed another rhythm. As we saw in earlier chapters, France was probably the first country to recognize the importance of regional problems in their modern guise, but planning did not become part of the French political culture until the 1940s (Bouchet, 1962, pp. 13–26). The first signs of official interest seem to have appeared before the War with the 1922 identification of nineteen 'economic regions', which were ascribed specific economic characteristics and a certain 'moral authority'. Things were laid to rest then, however, until the infamous Vichy regime proposed an administrative regionalization of the country—a project which was aborted with Nazi occupation (see Lagarde, 1977, pp. 31–33).

As has already been suggested, 1947 probably marks the birth pangs of French regional planning: *Paris et le Désert Français* set out the core/periphery problem in forceful language and called for urgent government action (Gravier, 1947). Starting with the *Comités d'Expansion Economique Régionale* in 1954 (an explicit copy of the *Centre d'Etudes et de Liaison des Intérêts Bretons*), steady increases in emphasis were given to regional matters, both from the standpoint of local administration and national economic planning (Lagarde, 1977, pp. 35–36). A first regional plan was produced for Brittany in 1956 (Ross and Cohen, 1975, p. 734). Decrees in 1959–1960 created twenty-one obligatory programming regions, and later actions in 1961 and 1963 created an interministerial coordinating committee and the now-famous *Délégation à l'Aménagement du Territoire et à l'Action Régionale* (DATAR) (see Hansen, 1968).

As elsewhere, Paris-initiated 'decentralization' took on a new urgency by the mid-1960s, with the establishment of a new administrative structure, serious industrial location controls, and the *métropoles d'équilibre* growth centres policy between 1964 and 1968. Within the framework of the Fifth Plan (1966–1969) these measures were in effect when the riots of 1968 shook France to its very foundations and the 'regional referendum' of 1969 finally toppled Charles de Gaulle and the Fifth Republic.[2]

There can be little doubt that by the end of the 1960s regional planning had become an idea in good currency once again (Friedmann and Weaver, 1979, pp. 1–2). Exact figures are difficult to obtain, but rough estimates of explicit regional policy expenditures in several industrialized countries at the turn of the decade are given in Table 6.1. Besides France, Britain, and the United States, data are also included for the Netherlands and Sweden, sometimes mentioned among the leaders in regional planning, and the Federal Republic of Germany, which suffered comparatively few regional problems (Perrin, 1974,

Table 6.1 Regional policy expenditures in selected countries, 1969*
(US millions of dollars)

Country	Regional policy expenditures	Gross domestic product	RPE GDP (%)	National government (current) disbursements	RPE NGD (%)	RPE per capita
France	90–108	134,748	0.07–0.08	47,039	0.19–0.23	1.8–2.2
Germany	75–125	152,089	0.05–0.08	48,762	0.15–0.26	1.2–2.1
Netherlands	20	28,067	0.07	10,716	0.19	1.6
Sweden	45	29,665	0.15	10,611	0.42	5.7
UK	552	110,307	0.50	36,598	1.50	9.9
US	409	934,346	0.04	269,157	0.15	2.0

* For sources see note 3.

pp. 175–202; Planque, 1977a).[3] Looking at per capita expenditure, most countries were spending in the vicinity of two dollars annually, although larger commitments were made by the United Kingdom and Sweden (at six and ten dollars, respectively). Britain also led everyone, including the United States, in total regional outlay, spending over half a billion dollars.

Not much can be made from such crude statistics, but several general observations seem in order. First of all, despite differing levels of public rhetoric and varying degrees of innovativeness there was very little difference in per capita expenditure between (most) countries with well-publicized regional planning efforts (e.g. France and the Netherlands) and those with less-elaborate, well-integrated programmes (e.g. the United States and Germany). It is impossible to be precise, but it seems probable that Germany spent as much per head as France, and the United States may have out-paced both Paris and the Hague. While negative controls have to be weighed in as well when attempting to assess a country's commitment to regional planning, comparing the 'carrot'—the financial incentive—does not provide many striking differences.

Turning secondly to absolute spending levels, only the United Kingdom and the United States, with four times Britain's population and several multiples its land area, spent a relatively disproportionate sum of money. Yet here again, if these expenditures are compared with total government outlays or the output of the country's domestic economy, they shrink to near insignificance. Even the UK's expenditures amounted to only 0.04 per cent. of the gross domestic product and less than two-tenths of 1 per cent. of London's current expenditures budget. Looking at these data cross-sectionally, variations from country to country were slight, the range running from around 0.15 to 0.26 per cent. of government spending, except for Britain and Sweden.

Attempting to assess the meaning of these numbers, there can be little doubt that regional economic policies had become a favoured topic of political discussion and academic theorizing in many countries by the end of the 1960s, but no serious financial commitment was forthcoming. Laws were passed, agencies set up, planners hired, and plans made *but* very little money was spent. Finally, the most important commitment by any of the measures presented was made by Great Britain, which during the 1960s had by far the longest-standing, best-financed programme.

2. THE 1970s IN AMERICA: CHANGE, EVALUATION, AND RETRENCHMENT

With the new decade things took a different turn. Richard Nixon was President of the United States, and after a two- or three-year period of transition his ideological antagonism towards federal spending for regional development was made a matter of public record. In 1971 he had attempted to defund both Appalachia and the Title V Commissions (Derthick, 1974b, pp. 106 and 119) and by 1974 he proposed to Congress that the Economic Development Administration, the seat of the Economic Development Centres (growth centres) programme, be abolished (Sundquist, 1975, p. 239). Thanks to the near immortality of existing public agencies this never actually took place, but funding stagnated for explicit regional programmes and emphasis shifted to block grant 'revenue sharing' and state-initiated rural development (Hansen, 1974b, pp. 296–303).

All 'community and regional development' spending by Washington moved up from \$0.2 to 9.9 billion between 1960 and 1978, creeping from 0.2 to 2.1 per cent. of the national budget (see Table 6.2). But the 'area and regional development' component of these outlays actually tapered off after Nixon's initial attack on the ARC and Title Vs, funding slipping from \$1.4 billion in 1972 to a low of \$0.9 billion in 1975. It was not until the first Carter administration budgets that spending worked its way up to 4 billion in 1978 (see Table 6.3), but this tripling of supposed regional policy expenditures was, in fact, an artifact of anti-cyclical fiscal policy.[4] As we will see below, a similar transformation was occurring in France and Britain as well by the mid-1970s.

Evaluations of the major US regional programmes vary, but few would argue that Nixon's supposed fiscal conservatism spelled the end for a particularly successful policy effort. EDA's own evaluation of its growth centre programme (1972) was very critical:

> EDA's experience in funding projects in economic development centres has not yet proven that the growth centre strategy outlined in the Agency's legislation and clarified in EDA policy statements is workable. The Agency's approach to assisting distressed areas through projects in growth centres has resulted in minimal employment and service benefits to residents of depressed counties (US Department of Commerce, 1972 p. v.).

In fact, only a few authors, who had helped develop and justify the growth centres approach in the American context, gave the concept continuing support (e.g. see Berry, 1973, pp. 353–355; and Hansen, 1974b, p. 295; 1975, p. 150).

Table 6.2 US Federal budget outlays by function, 1960–1978 (Billions of Dollars) (Taken from US Department of Commerce, Bureau of the Census, 1978, p. 260)

Category	1960	Total (%)	1965	Total (%)	1970	Total (%)	1975	Total (%)	1978	Total (%)	Change 1960–78 (%)	Change index*
							Year					
1. National defense	45.2	49.0	48.6	41.0	79.3	40.3	85.5	26.2	107.6	23.3	138.1	0.34
2. Human resources	25.5	27.7	35.4	29.9	72.7	37.0	168.7	51.7	283.3	51.6	834.5	2.08
3. Income security	18.3	19.8	25.7	21.7	43.1	21.9	108.6	33.3	147.6	31.9	706.6	1.76
4. Education training, employment and social security	1.0	1.1	2.1	1.8	7.9	4.0	15.8	4.8	27.5	5.9	2650.0	6.60
5. Other non-defense	21.4	23.2	34.4	29.1	44.8	22.8	71.9	22.0	116.2	25.1	443.0	1.10
6. Community and regional development	0.2	0.2	1.1	0.9	3.2	1.6	3.7	1.1	9.7	2.1	4750.0	11.84
7. Revenue sharing	0.2	0.2	0.2	0.2	0.5	0.3	7.2	2.2	9.9	2.1	4850.0	12.09
8. Total outlays	92.2	—	118.4	—	196.6	—	326.6	—	462.2	—	401.3	—

* Change index = percentage change 1960–1978 in each category divided by the percentage change 1960–1978 for total outlays.

Table 6.3   US Federal outlays by detailed function, 1972–1978 (Billions of Dollars) (Taken from US Department of Commerce, Bureau of the Census, 1975, p. 227, and 1978, p. 261)

Category	Year							Total (%)	Change 1972–1978 (%)
	1972	1973	1974	1975	1976	1977	1978		
1. Community and regional development	4.7	5.9	4.9	4.4	5.3	6.2*	9.7	49.7	106.4
2. Community development	3.1	3.1	3.0	3.1	3.5	3.5	4.0	20.5	29.0
3. Area and regional development	1.4	1.4	1.1	0.9	1.3	2.1	4.0	20.5	185.7
4. Disaster relief	0.4	1.6	0.8	0.4	0.5	0.6	1.7	8.7	325.0
5. General revenue sharing	—	6.6	6.1	6.1	6.2	6.8	6.8	34.9	—
6. Other general purpose fiscal assistance	—	—	0.6	0.9	0.9	2.7	3.0	15.4	—
7. Total	—	—	11.6	11.4	12.4	15.7	19.5	—	—
8. Area and regional development as percentage of total	—	—	9.5	7.9	10.5	13.4	20.5	—	—
9. Area and regional development as percentage of community and regional development	29.8	23.7	22.4	20.5	24.5	33.9	41.2	—	—

* Adjusted to add to calculated subtotal.

PWEDA's other major regional programme, the Title V state/federal regional action commissions, has also received extremely critical evaluations (e.g. Derthick, 1974; Hansen, 1975; US Department of Commerce, 1974a). As mentioned earlier, Title V was basically meant to buy congressional approval of the Appalachian Regional Commission and it never achieved presidential support, thus never receiving a level of funding equivalent to that of the ARC. ARC funnelled over $5 billion in federal funds into its thirteen member states from 1965 through 1979, for an average of more than $330 million annually, while during fiscal year 1979, for instance, all ten existing Title V Commissions had only $63 million to divide among themselves (Sinclair, 1979). The Carter administration's 1980 budget figure was $74 million (*APA News*, March 1979).

Because of its lack of financial leverage Title V never gained the backing of most of the state governors involved, and therefore had very little impact. 'Action plans' were prepared and money was spent on often questionable technical assistance programmes and limited low-level hardware investment. The Ozarks Regional Commission, for instance, spent comparatively large sums of money on development of a regionwide computerized information system which was never fully operationalized and proved all but useless.

Turning to the most important of the American programmes, ARC, some of the better-known evaluations have been kinder to it than EDA efforts (e.g. ACIR, 1972; Newman, 1972; Rothblatt, 1971; US Department of Commerce, Economic Development Administration, 1974b). Hansen (1975, p. 147) suggested that this was primarily because ARC had a broader scope than merely attempting to alleviate unequal income and employment opportunities, as EDA had concentrated on doing. Derthick (1974) argued, though, that:

> The commission's self-evaluations, done with competence and a candor rare in official documents, do not claim that the commission's programmes have had much effect on the region's economic performance. . . .
> [It] concedes that anyone who expected a 'regional solution' to the area's problems will doubtless be disappointed, but argues that the failure of 'bureaucratic solutions' to Appalachia's problems justifies an organization that gives states and their governors 'a policy voice in the management of the federal system.' This remains commission doctrine; the staff continues to hope that in time it will be possible to demonstrate the practical merits (Derthick, 1974, pp. 105–106).

In one of the last evaluations of the Johnson administration regional policies by a leading regional scientist, Hansen (1976) concluded:

> . . . it would appear that regional development processes, as well as the impact of federal outlays upon them, are very imperfectly understood. Questions of the relationship between city size and economic efficiency also are far from being resolved. . . . Although survey results indicate that concern about human settlement patterns is widespread, almost nothing is known about the priorities people attach to spatial distribution issues in relation to other social and economic

problems. Moreover, it may be more advisable to attack many problems directly rather than by trying to alter the sizes of places. . . .

   Given the difficulties outlined here, it is highly questionable whether policy makers should attempt to determine where people should live or where economic activities should be located (Hansen, 1976, pp. 23–24).

Hansen went on rather incongruously to suggest founding a White House-based Regional Development Agency, as a 'learning experience'. The Carter administration, in fact, proposed formation of a not-dissimilar umbrella agency as part of its promised administrative reorganization—the Department of Development Assistance (DDA). However, neither the DDA nor its related Department of Natural Resources curried much favour with the established line agencies or their defenders in Congress.

   As an outgrowth of the sunbelt/snowbelt issue (e.g. see Moynihan, 1978; 'Special feature', 1979; Sternleib and Hughes, 1977; Watkins, 1979), Congress and the administration debated whether to extend the network of Title V Commissions across the rest of the country, as well as consolidating economic assistance programmes originated under Titles I, IV, and IX of the Economic Development Act. This would have expanded EDA's budget to $1.3 billion (Popper, 1979). Various Senate and House versions of the measure suggested that Title V's coverage should be extended to at least some part of all states not currently involved in the programme. The practical importance of this apparent renewed interest in regional policy was nil, however. At the time of the Carter administration's defeat at the polls both ARC and the Title V Commissions were operating on emergency appropriations (*APA News*, September 1980), and in the last days of the 96th Congress the Administration was finally given $624 million (*APA News*, February 1981).

   Following the example of the Thatcher government in Britain, discussed below, the conservative Reagan administration attempted to axe EDA to $48 million in fiscal 1982 and scheduled it to go out of existence in 1983 (see Kashdan, 1981a). Congress, however, decided to reauthorize EDA at $290 million for 1982 (*APA News*, September 1981), protecting temporarily a popular slush fund. Reagan is also making a concerted effort to shelve the ARC as part of his general 'budget cutting' campaign. In their stead has been suggested the British notion of 'reindustrialization' through 'enterprise zones', discussed below (see *APA News*, March 1981; Butler, 1981; Friedman and Schweke, 1981; Golob, 1983; Grossman, 1979; Hall, 1981a, 1981b; Kashdan, 1981a; Sternleib and Listokin, 1981). Pending legislation would permit the US Department of Housing and Urban Development (HUD) to begin experimenting with the new strategy, initially, through creation of ten to twenty-five such areas (*APA News*, July 1981, August 1981; Butler, 1982; USGAO, 1982; White House, 1982a, 1982b). Though slow in getting through Congress, some version of the legislation is expected during the 1984 session, and almost a dozen states have already enacted their own enterprise zone laws (Gold, 1982; National Urban League, 1982; Sidor, 1982; US HUD, 1982).

3. THE 1970s IN EUROPE: 'REGIONAL POLICY FOREVER?'

In France and Britain developments continued to move in a more linear direction until the general economic and fiscal crisis of the mid-1970s, and there were remarkable similarities on both sides of the Channel. Growth centres programmes retreated into the background, as industrial relocation subsidies and controls were elaborated into a maze of confusing geographic complexity. Britain was divided up into four different types of assisted areas and France into six (Sundquist, 1975, pp. 65 and 128).[5] Both countries became increasingly aware of the problem of decentralizing tertiary and quaternary economic activities—meaning services, office work, and research and development. While claims were made that secondary industry was being effectively urged to move into outlying areas from London and Paris, it was recognized that the 'leading economic sectors', the ones which were acting as the real motor for national economic growth, were becoming even more concentrated in the national core areas; and they responded *very* poorly to location controls (Alden and Morgan, 1974, pp. 82–121; Cameron, 1974, pp. 65–102; Law, 1980; Pickvance, 1980; Prud'homme, 1974, pp. 33–63; Sant, 1975; Sundquist, 1975, pp. 76–84).

As had been suggested by leading theorists, the very nature of services, management, and research activities seemed to require frequent face-to-face contacts, both in the public and private sectors, and it proved extremely difficult to move them around (Daniels, 1969, 1977; Goddard, 1973; Thorngren, 1970; Tornqvist, 1970). Not only have multiplant and multinational enterprises followed locational strategies which distribute different activities according to their role in the production process but the economic linkages of decentralized assembly operations have also tended to remain minimal, casting a growing doubt on the potential effectiveness of regional policies to bring about substantially balanced regional growth and development.

It is difficult to obtain comparative time-series and cross-sectional information on the proportion of the labour force engaged in various stages of the production process, but Table 6.4 presents sectoral changes in the composition of gross domestic product for France, the United Kingdom, and the United States during the period 1960–1975. These data, covering the approximate life-span of the regional policy approach to planning, document convincingly the ever-increasing importance of service activities to industrial expansion: the 'tertiarization' of society (Lipietz, 1979; Marquand, 1979). The dynamic sectors of the economy over the last two decades, what Perroux called *les industries motrices*, have been the services—the very branches of productive activity which have proven so stiffly resistant to government location incentives. As shown in the table, public utilities and various personal and industrial services (e.g. lines 4, 7, 8, 9, and 10) have become predominant in all three countries, both in terms of contribution to the GDP and rates of growth.

By the mid-1970s a third generation of academic textbooks had appeared dealing with planning and regional economic development. These books set

Table 6.4 Changes in industrial structure 1960–1975*: France, United Kingdom and United States (Taken from UN Statistical Office, 1977, pp. 277, 1107–1109, and 1154)

Category	France			United Kingdom			United States		
	GDP 1960 (%)	GDP 1975 (%)	Change index	GDP 1960 (%)	GDP 1975 (%)	Change index	GDP 1960 (%)	GDP 1975 (%)	Change index
1. Agriculture, hunting forestry, and fishing	6.8	5.6	0.31	3.5	2.5	0.96	3.8	2.7	0.21
2. Mining and quarrying	1.7	—	—	2.6	1.6	—	2.0	1.6	0.49
3. Manufacturing	35.0	31.3	0.71	32.1	26.0	1.85	24.7	24.0	0.92
4. Electricity, gas, and water	1.6	—	—	2.4	2.8	2.71	2.0	2.3	1.47
5. Construction	8.0	6.9	0.79	5.6	6.2	0.55	5.7	3.8	0.13
6. Material production	56.3	43.8	0.65	46.3	39.0	—	38.2	34.4	0.74
7. Wholesale and retail trade, restaurants and hotels	11.5	—	1.33	10.7	8.9	0.94	17.0	18.6	1.24
8. Transport, storage, and communication	4.9	—	1.22	7.7	8.3	1.33	5.8	6.9	1.51
9. Finance, insurance, estate, and business services	4.4	40.1	1.47	8.9	12.8	—	16.4	18.1	1.29
10. Community, social, and personal services	10.5	—	—	8.3	9.0	—	6.0	7.2	1.50
11. All services	31.4	40.1	1.44	35.6	39.0	1.32	45.2	50.8	1.32

* 'Old' (1960) and 'new' (1975) systems of national accounts are not strictly comparable. Change index represents the percentage change 1960–1975 in each sector divided by the percentage change 1960–1975 of GDP.

out fairly tight, systematic discussions of regional planning, from its theoretical bases to its historical development, contemporary programmes and institutional structure, and, finally, some evaluation of its recent performance. Not surprisingly most of these works dealt primarily with Britain, giving lesser coverage to the United States, France, and other countries. Notable among them were Jeremy Alden and Robert Morgan's *Regional Planning: A Comprehensive View* (1974), John Glasson's *An Introduction to Regional Planning* (1974), Peter Hall's *Urban and Regional Planning* (1975), and Jean-Claude Perrin's *Le Dévelopment Régional* (1975).[6]

The assessment offered in these sources as to the success of regional planning (from approximately 1965 through 1972 and primarily in the United Kingdom) might be described as something between guarded optimism and guarded pessimism. Glasson's concluding remarks in summarizing his discussion of regional planning (interregional planning in his vocabulary) is representative:

> Yet in spite of this impressive array of measures, UK inter-regional planning has failed as yet to achieve anything more than a holding operation. Of course, this in itself can rightly be considered to represent some degree of success but there is still a long way to go. The process is long-drawn-out and premature relaxation can undo much good work (Glasson, 1974, p. 211).

Pertaining to France, Perrin's conclusions were even less optimistic:

> Thus, neither the interventions of DATAR, nor the regionalization of infrastructure programmes (reform of 1964), nor the Regional Plans for Economic Development (PRDE) established by the regional prefects in accordance with the Sixth Plan have permitted the French regions concerned to close in a significant fashion on the self-development goals set by the National Commission for Territorial Planning (CNAT) (Perrin, 1975, p. 201, my translation).

More recent mainstream evaluations of regional policy performance in Britain and France has maintained much the same tone (see, for example, Begg and Lythe, 1977; *Cahiers*, 1976; Law, 1980; MacLennan and Parr, 1979; Planque, 1977b; Manners *et al.*, 1980; Richardson, 1978a; Roland, 1977).

4. THE DEATH OF REGIONAL POLICY

If the retrenchment from regional planning during the early 1970s was the most evident in the United States, where Great Society programmes were either transformed or dismantled, it also became plainly apparent in France and the United Kingdom by the end of the decade. In a series of discussions with the author during the spring of 1978, Jean-Claude Perrin, Director of the Centre for Regional Economics at the University of Aix-Marseille, argued that, effectively, France had no current regional economic policy, let alone integrated regional economic and physical planning.[7] This view corresponds closely with published findings (e.g. Delors, 1978; Prats, 1979; Ross and Cohen, 1975, pp. 727–750).

On the national level, the extreme de-emphasis of regional problems by President Giscard's alliance was in open display during the 1978 parliamentary elections, as the Socialist–Communist coalition was allowed to monopolize the topic as a campaign issue. This tendency was perhaps best reflected by the conceptual framework developed during preparation of the Seventh Plan. In the post-energy embargo days of the mid-1970s it was argued that short-term battles against differential regional unemployment should become the centre-piece of national policy (Passeron, 1979). A situation of slow growth and recession was being aggravated by the rationalization of industrial location patterns and the subsequent need for decentralization of tertiary activities. Regional inequalities, which had fallen between 1962 and 1970, could be expected to remain unchanged—meaning they were still extremely pronounced—in 1980 (p. 168). Rather than considering how best to restructure lagging regional economies, encouraging higher equality through new growth, all that could now be hoped for was a holding action—fighting brush fires on a case-by-case basis (pp. 165–166). Expectations would have to be lowered. Like the US regional programmes discussed earlier, French regional policy would have to be transformed into an adjunct to the government's tool kit of anti-cyclical measures. While the still-born eighth Plan devoted more attention to regional issues than did its predecessor, because of a change in governments it was never put into effect.

This change took place in May 1981, when the Socialist, François Mitter-rand, was elected President of the French Republic. After the second round of legislative elections in June he was given a Socialist majority in the National Assembly as well. This new Leftist government, the first in thirty-five years, called for a decidedly different approach to regional problems: two of the main planks of its election platform were nationalization of key branches of industry and decentralization of the French political structure. A ninth Plan is currently in preparation and should be in place by 1985. The new Minister of Planning, Michel Rocard, is noted for his commitment to 'self-management through decentralization . . . [and] cooperative and associative movements' (Le Monde, 1981, p. 27), recalling the Proudhonist traditions discussed here in Chapter 3.

Because of the radical change implied by such strategies, it will be some time before their real significance for regional development and planning can be assessed. Gaston Defferre, Minister of the Interior and Decentralization, has said that regional planning and DATAR will both be affected, but emphasized that political construction of a new 'Girondine' France must come first (Le Monde, 1981, p. 45). So far implementation has been sluggish, at best. In discussions with Professor Remy Prud'homme, at the Massachusetts Institute of Technology, during January and February, 1983, he argued that little of substance had been achieved. Despite the enactment of numerous 'decentral-ization' measures—semi-autonomous status for Corsica, election of regional prefects, establishment of regional budgetary councils, etc.—financial control remains in Paris. Perhaps most importantly, after two years in office, socialist macro-economic policies went through a 180 degree turn, coming to resemble closely the approach favoured under the Giscard government's 'Plan Barr'. Regional planning apparently remains on hold.

In Britain also the long-standing consensus as to the role of regional policy in coping with national economic problems has come into question (Damesick, 1982; McCallum, 1979). Beginning as early as 1970, the Conservative government, brought to power in the June general elections, began to relax industrial and office development controls which had been meant to steer new growth away from the South East. This was the first move in a more general attempt to 'disengage from national and regional economic policy' (McCallum, 1979, pp. 20–21). Coming at almost the same moment that Nixon began defunding American regional planning efforts, these tentative moves were stalled, however, by the worsening recession in Britain in 1971–1972. By August 1972, with passage of the new Industry Act, the Heath government found itself readopting in general form the same grant system and assistance machinery which had been initiated by Labour in the mid-1960s. Significantly, any serious attempt to control the growth of tertiary activities (office space) in London was abandoned. This apparent waffling on regional policy measures by Conservative and Labour governments alike caused the Trade and Industry Subcommittee of the House of Commons Expenditure Committee to conclude, after a fourteen month review, that: 'Much has been spent and much may well have been wasted. Regional policy has been empiricism run mad, a game of hit-and-miss, played with more enthusiasm than success' (Great Britain, 1973, p. 73, quoted in McCallum, 1979, p. 24, and Manners et al., 1980, p. 67). From another perspective, though, it might be argued that regional policy in Britain, as in the United States and France, was well on its way to becoming primarily an adjunct to the State's bag of short-term, anti-cyclical policies and was used quite explicitly—if covertly—to field the political flack stemming from fluctuating regional and national unemployment trends.[8]

Faced with one of the most compromising structural situations in the industrial world, Britain was hard hit by the 1973 oil embargo. Coming in the train of EEC entry, a steadily deteriorating balance of payments, growing structural unemployment, and an embarrassing inability to compete on international markets, Labour found itself back in power in March 1974. There followed three major swings in regional policy: (1) traditional measures such as unemployment premiums, assistance grants, and office location controls were at first strengthened or reinstated, and a European Regional Development Fund was established (March 1975); then (2) local government was reorganized, the National Enterprise Board was set up, followed by the Scottish and Welsh Development Agencies; and, finally (3), under heavy pressure from the IMF to deal with stagflation, controls were relaxed again, a programme of seeking national 'sectoral efficiency' was announced, and general public spending cutbacks ate badly into all manner of regional policy spending.[9] When the minority Callaghan government failed in its March 1979 bid for Scottish and Welsh devolution, the government itself lost its necessary coalition of supporters and was turned out of office two months later. While there were obviously other underlying reasons for discontent, it is important to note that the government fell on the basis of its 'regional policy' failure (Dunford, Geddes, and Perrons, 1980).

Regional programme cuts forced by circumstances upon the outgoing Labour administration have been pushed with enthusiasm by the Conservative Thatcher government, which has taken a direct approach to axing regional expenditures. Townsend has outlined the major changes:

> The new Conservative government, elected on 3 May 1979, had by August abolished the English Regional Economic Planning Councils and Office Development Permits (ODPs), reduced the Civil Service office dispersal programme and the development control powers of the County Planning Departments, and initiated reviews, which produced fresh decisions by October, of Inner Cities policy and the future of QUANGOs (Quasi-Autonomous Non-Government Organizations, such as New Town Development Corporations). Among the most prominent changes was a reduction in the financial and geographical scales of regional aid to industry, announced by Sir Keith Joseph, Secretary of State for Industry, on 17 July. We might have expected reversions of policy towards the position under former Conservative administrations; but if we note that Planning Councils, ODPs and many QUANGOs survived from the 1964–70 Labour administration, then it may be that the parallels lie more in the early 1960s than 1970–74 (Townsend, 1980, p. 9).

Geographic changes in assistance are shown on Figures 6.1 and 6.2. Figure 6.1 portrays assisted areas as of June 1979, which, substantially, had remained unchanged since 1972 (see note 4). Figure 6.2 shows the reductions phased in through August 1982. Special development areas, the worst problem regions, were left basically intact, but development areas and intermediate areas have been significantly reduced. In all, the percentage of the British working population covered by assistance has dropped from 40 down to 25 per cent. As well, the restrictions on plant expansion imposed by the Industrial Development Certificate system—the 'stick' side of regional policy—has been effectively eliminated (Townsend, 1980, pp. 10–11).

The trajectory of British economic development policy has also been reoriented by the creation of a number of urban 'enterprise zones' (EZs) (*Built Environment*, 1981), which have served in part as a model for the American initiatives mentioned earlier.[10] In 1977, Peter Hall, at the University of Reading, observed that, given structural changes in the international economy, strategies meant to inject science-based technology and service-oriented industries into the inner-city areas of declining industrial centres such as Manchester and Glasgow seemed futile, especially in an era of fiscal stringency. He went on to suggest that the establishment of a number of small-scale, experimental 'freeports', where most governmental legislation and regulations could be waived, might rekindle in Britain the entrepreneurial spirit of innovation and risk-taking which currently characterize the newly developing countries of Asia (Hall, 1977). Cutting the red-tape and planning regulations resulting from years of increasing government intervention, he argued, could perhaps point

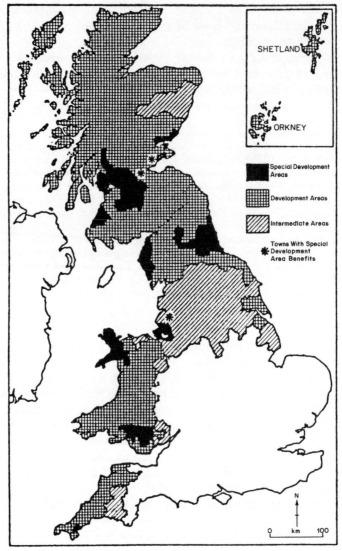

Figure 6.1   Assisted areas in Britain (June 1979). (Taken from Townsend, 1980, p. 10)

the way towards self-sustaining economic growth, based on the free play of market forces. This would once again build up the capitalist economy—from the bottom up.

Six months later, Sir Geoffrey Howe, then a senior member of the Conservative Opposition, drew upon Hall's idea to propose a model of 'pure' enterprise zones, laying stress almost exclusively on freeing the private sector from government interference (see Howe, 1981). When the Conservative Party came to power in 1979 Howe became Chancellor of the Exchequer. In his first

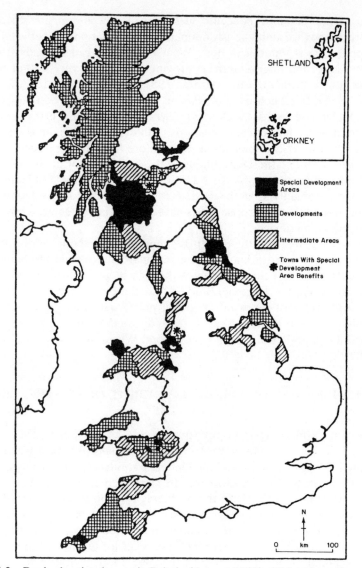

Figure 6.2   Revised assisted areas in Britain (August 1982). (Taken from Townsend, 1980, p. 12)

budget, introduced in March 1980, it was announced that an Enterprise Programme which called for the initial establishment of five EZs of approximately 500 acres each was to be pursued. By November 1980 legislation was in place, and by the summer of 1981 a total of eleven enterprise zones were operating around the United Kingdom: one each in Scotland, Wales, and Northern Ireland, with the rest distributed around England. Another round of designations took place in the spring of 1983.

Like earlier planning concepts such as new towns and growth centres, enterprise zones are ambiguously defined, and the relationship between theory and application is tenuous at best. What began as a call for radical market economy and 'non-intervention' is proving in practice yet another form of bureaucratic incrementalism (Barnes, 1982; Hardison, 1981; Pirie, 1982; Taylor, 1981), the functioning of which is based in large part on public sector funds (McDonald and Howick 1981; Tym et al., 1982). Further confusion stems from most theoretical discussions unfolding in the realm of 'economic policy', while practical applications seem more like urban physical renewal, under the control of the Department of the Environment in Britain (and its American counterpart across the Atlantic, the US Department of Housing and Urban Development). Like the garden cities idea at the turn of the century, it could be argued that enterprise zones have already been neutered into a 'safe' physical planning device.

However this may be, the enterprise zone idea—at the level of doctrine—probably sounds a death knell for the regional-science style of planning discussed in the last two chapters. Given the lack of government interest in the United Kingdom, the Regional Studies Association has recently called for a serious rethinking of the whole regional policy question, promising a new, privately prepared 'Barlow Report' (see *Regional Studies*, 1982). However, how are the experiences of the post-Second World War period to be interpreted?

5. THE DILEMMAS OF LIBERALISM: RIGHT OF CENTRE, LEFT OF CENTRE

Can we find a more synoptic interpretation of the regional policy changes discussed in the preceding pages? From one perspective, post-war regional planning in France, Britain, and the United States has been a problem-solving mechanism for central government. In simplified form, this has been an attempt to come to grips with the problems of geographically concentrated unemployment and low family incomes. As such, it has been a politically responsive *place-oriented* welfare policy, but, more importantly, through its emphasis on infrastructure investment, tax relief for business, and direct subsidies to capital, regional planning has become a tool of State intervention in the process of structural economic change. Although the explicit goals of regional planning to bring relief to certain micro-areas (e.g. the declining mining valleys of Wales, the Ozarks, the Scottish Highlands, Lorraine-Nord, or Appalachia) have manifestly 'failed', regional policy has helped to rationalize the spatial structure of activities in broader areas (e.g. Southern Wales, the southern Midwest, Scotland, 'Lotheringia', and the Atlantic Seaboard) (Damette, 1980; Geddes, 1979; Morgan, 1979).

In an effort to legitimize apparent failures and the social hardships of rationalization, central governments have tried to establish local identification with their policies through *regional institution building*. Institutions such as the

Title V Commissions, the Welsh and Scottish Development Agencies, the Appalachian Regional Commission, and the Highlands and Islands Development Board, which delegate some limited measure of authority to regional decision makers, not only create local identity with planning activities but *local responsibility as well*. This all goes towards building and maintaining what Rees and Lambert (1979) have referred to as a regionalist consensus. A next logical step after regionalist consensus, in the face of ambiguous and contradictory results, is some form of further devolution, as attempted by the British Labour government in March 1979 and the Socialist government in France with its 'decentralization' programme. Another complementary tactic is the post-Keynesian retreat from now traditional government welfare responsibilities, being initiated today by decision makers in all three countries. The part played by regional planning among government managerial strategies has been contextuated by evolving economic and fiscal circumstances. At each step of the way, regional planning has been reoriented to aid in attempts to buoy up the capitalist economy. First it served as a vehicle for physical reconstruction; then it acted as a conduit for social spending, keeping the peace in face of relentless economic rationalization; finally, having served its purpose, with the trend towards devolution and government austerity, by the end of the 1970s regional policy had become an early casualty of selective budget cuts. It was an 'unaffordable luxury' in an era of national and international economic crisis (Passeron, 1979; Richardson, 1978a; Wilson, 1982).

NOTES

1. Martin and Leone (1977, pp. 19–39) go through the legislative history of this period, and Rothblatt (1971, pp. 37–65), focusing on the emergence of the Appalachian Regional Development Act, gives a running account of attempts to formulate regional policy legislation from 1954 to 1965.
2. The referendum read: 'Do you approve of the referendum ("projet de loi") submitted to the French people by the President of the Republic, relative to the creation of regions and the renovation of the Senate?' It was defeated by 53.2 per cent. of the French voters that went to the polls (see Lagarde, 1977, pp. 43–48).
3. Comparable data are extremely difficult to obtain. European figures were taken from the original OECD (1970) report on regional policy. American data were estimated from statistics cited in Hansen (1975). The questionable reliability of all such estimates is demonstrated by the profound variations reported for the same types of expenditures in the same countries by OECD in its revised 1974 publication *Re-appraisal of Regional Policies in OECD Countries* (pp. 119–121). The closest annual average statistics given for the subject countries in millions of US dollars were as follows: France (1970–1972) $207.0; Germany (1967–1970) $54.7; Netherlands (1967–1971) $38.6; Sweden (1967–1968) $43.8; and United Kingdom (1967–1968) $326.9, (1971–1972) $579.4. A comparison with Table 6.1 will show that for the nearest available year, the only country for which the estimates were even similar was Sweden ($45 versus $44). (By jumping ahead two years the British estimates also get closer, i.e. $552 versus $579.) The most obvious reason for these discrepancies is undoubtedly definitional.
   Statistics for Italy were purposely deleted from consideration here, because it was felt that they would introduce an even greater margin of confusion. Italy spends by

far and away the most money of any Western country on 'regional policy'; according to the two OECD sources, respectively, in millions of US dollars (1969), $750 or $13.9 per capita, and annual averages for 1965–1970 of $745.6 and 1971–1975 of $2,448.0. This regional programme is of a distinctly different kind from the others under consideration, looking back to the comprehensive TVA-type effort of the 1930s, only this time encompassing half a country (see OECD, 1976, pp. 29–57). As noted by the OECD (1974 and 1976) the Italian programme is not a palliative measure, but has aimed at achieving profound economic structural change at the national level.

For this reason, substantial amounts of public investment have taken place which have no comparison in other mixed industrial economies. Further discussion of the Italian situation is available in Hansen (1974a), Sundquist (1975), Holland (1976a), Dunford (1977), Mingione (1977), Rogers (1979), Arcangeli, Borzaga, and Goglio (1980), and Arcangeli (1982).

4. Funding levels are not self-evident indications of regional policy commitments. The Economic Development Administration's budget varied widely during its first fourteen years, ranging from a low of around $300 million to a high of almost $3 billion during the recession period of 1975–1977. But the way money has been allocated among various titles of the EDA's programme and the real emphasis of this spending has changed the whole thrust of the EDA effort. The Administration started as a non-metropolitan-oriented development agency which gave grants to public bodies and loans to private business in order to stimulate depressed areas with high levels of unemployment. The language of its original legislation made it extremely difficult to spend such money in metropolitan areas, and EDA's early efforts were focused on long-term 'structural' investments, some in experimental Economic Development Centres.

With Nixon's attempt to kill the EDA it found a host of new supporters in Congress. To some extent this was because it had become the nation's most blatant 'pork barrel'; it was rumoured that congressmen were its only constituency. This political patronage function, as well as post-oil embargo stagflation, was used to entirely reorient the Administration's functions. After 1974 most of its activities were focused on combating cyclical unemployment, and the cornerstone of EDA spending has become accelerated public works programmes, largely in urban areas. The Carter administration's 1980 budget cut EDA funding by 4 per cent. from the 1979 figure, to $566 million. Of this, $150 million were allocated to the Inland Energy Impact Assistance Program, an emergency trouble-shooting effort to mitigate the impacts caused by boomtown energy development in smaller communities. There was also a 39 per cent. increase in EDA's overheads for staffing three new regional offices (*APA News*, March 1979). This pattern continued at a $117 million higher level in 1981 (*APA News*, February 1981). Thus when the Reagan administration set EDA to be phased out in 1983 EDA expenditures only tangentially fit in the category *regional planning*, even though the administration worked under basically the same legislative mandate, applying basically the same investment incentives, by basically the same criteria.

A recent discussion of EDA is given in Martin and Leone (1977, pp. 45–91); also see the official history, 1965–1973, in the US Department of Commerce (1974a).

5. In Britain, as of 1972, the breakdown consisted of (1) special development areas, (2) development areas, (3) intermediate areas, and (4) derelict land clearance areas. There were also 'towns where development area benefits are available' and 'towns where special development area benefits are available', and, added to this, was the special category of 'Northern Ireland'.

In the same year, France had four zones of investment subsidies and controls, with a number of subcategories. Zone A included areas of maximum benefits, intermediate benefits, and lower benefits. Zones B and C concentrated on subsidies for relocation of tertiary activities, except where particularly severe unemployment problems qualified them for zone A-type benefits. And, finally, zone D, the Paris Basin and a broader peripheral ring surrounding it (except on the east), were divided, respectively, into zones of investment controls and no investment controls. For further details, see the White Paper (1972) and DATAR (1972).

6. Another series of publications appearing about this time presented narrative discussions of largely unrelated case studies. Some of the better-known examples drawn from the industrialized world (already cited above) included Derthick (1974), Hansen (1974a), Kuklinski (1975), and Sundquist (1975). These followed more closely in the empirical tradition of Hansen (1968), McCrone (1969), and Rodwin (1970). David Gillingwater's (Gillingwater, 1975; Gillingwater and Hart, 1978) theoretical approach to procedural problems in regional planning seems to be a unique effort to direct ideas developed in the planning theory literature towards regional policy making (see Friedmann, 1975).

7. I would like to express my most sincere appreciation to Jean-Claude Perrin and my other colleagues at CER during the spring of 1978 for their help and tireless willingness to respond to my questions. Special thanks are due to Bernard Planque, Pierre-Yves Leo, and Jean Philippe. I also owe a debt of gratitude to Katerhine Coit, formerly of the Moulin de Trois-Cornets in Paris. Needless to say, however, the generalizations presented here are my own and do not necessarily express the stated views of my collaborators or interviewees.

8. Law (1980, pp. 55–56) argues that this has always been the principal function of British regional policy.

9. For a more detailed account of these changes, see McCallum (1979, pp. 19–38), Law (1980, pp. 48–56 and 234–245), and Manners et al. (1980, pp. 33–68).

10. Special thanks to Don Golob for helping me keep abreast of the urban enterprise zone literature.

CHAPTER 7

Development, Underdevelopment, and Uneven Development: A Debate on the Left

As we saw in Chapter 6, disenchantment with the vintage regional policies of the 1960s has become very widespread. By the end of the 1970s most pretences of regional planning in France, Britain, and the United States had been shelved in favour of anti-cyclical policies and a new monetarism. Theorists have reacted to these changes with what Richardson (1978a) described as a 'growing pessimism'.

Other events have reinforced these tendencies. Failure of the first UN Development Decade and widespread violence in the West's major urban centres had spawned increasingly critical intellectual responses by the late 1960s. Step by step many practitioners in the fields of development and urban studies became radicalized (Castells, 1972; De Souza and Porter, 1974; Harvey, 1973; Lipton, 1977; McGee, 1971; Seers, 1977). Because of their direct links with regional science it was probably inevitable that this ferment should spill over into regional planning as well (Friedmann and Weaver, 1979).

Another factor of importance has been the advent of a new, militant regionalism in several Western countries. Along with Canada, Belgium, and Spain, France, Britain, and the United States have been in the forefront of this movement (Lafont, 1967; Nairn, 1977; Rafuse, 1977). The hegemony of the nation-state, which accompanied the rise of industrial capitalism, as we saw in Chapter 2, shows signs of weakening with the global spread of economic production and markets. As central governments have lost their leverage to deal with economic problems, past regional grievances, cultural differences, and struggles over economic resources have reasserted themselves.

However unwillingly, this has forced planners once again to explore the linkages between political radicalism and development. For the first time in several decades questions about rural/urban contradictions, class relations, and political sovereignty have entered the debate. The concerns of early regional planners and the precursors of regional planning have reappeared. In this chapter I will explore these developments and show how they are contributing to new approaches to regional theory.

112

1. UNDERDEVELOPMENT AND DEPENDENCY

The first model of regional planning was drawn from the experience of western Europe. The growth of national markets and major urban centres in the heartlands of Britain and France were obviously linked to the decline of local economies in peripheral regions. Development strategies were meant to reverse these trends by recreating viable regional communities.

The regional policies of the 1960s and '70s were modelled in large part on the economic history of North America. Planning strategies sought to bring about the development of a country's urban system based on resource exploitation and the subsequent expansion of industrial activities. From one perspective, growth centre strategies can be seen as an attempt to induce the kind of economic growth experienced in favoured regions of the United States following the Civil War.

Reflecting preoccupations of the present era, recent theories of regional *underdevelopment and dependency* owe much of their inspiration to interpretations of Third World experience. Writers like Frank (1967), Hymer (1968, 1972a), Sunkel (1970, 1973), Coraggio (1972, 1975), Emmanuel (1972), and Amin (1974) made a frontal attack on conventional development and spatial planning notions because of their apparent inconsistency with the recent economic history of Latin America and Africa. The central tenets of this line of thinking have come to be applied to regional problems in Western countries as an indictment of the multinational corporation and internal colonialism.

The Third World model of political economy

Underdevelopment/dependency theory (UDT) is probably the first truly *political* economic model to gain currency since the Second World War. In its more technical statements it shares much in common with the considerations of Smith, Ricardo, Marx, and Veblen. It is primarily concerned with fundamental macro-economic questions, such as the distribution of revenue, the accumulation of capital, the role of institutions, and the relation of these to economic growth. Beyond this, UDT also begins an analysis of the essential social and political relations of underdevelopment.

Figure 7.1 shows the main structural relationships postulated in the UDT framework. As with Myrdal's discussion of the development problem covered in the Chapter 5, UDT has an explicitly spatial component, with much of what is fundamental to the development process hinging on increasing geographic disparities. The *core area* in Figure 7.1 was meant by the original writers to represent the industrialized capitalist countries, like France, Britain, and the United States. Here the economic history of the last two hundred years had created a vast store of fixed capital, a skilled labour force, and relatively wealthy consumer markets with high levels of effective demand for a wide range of products. This market was said to have come about, as in neoclassical formulations, because of innovation, increasing capital intensity and labour

114

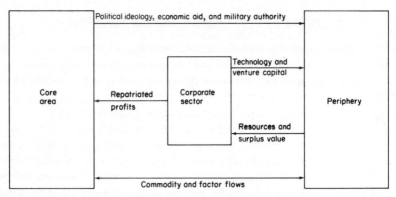

Figure 7.1   The structural relations of underdevelopment

productivity, and rising real wages—won in large measure through effective union organization. These same processes helped create ever-larger production units—firms with increasing levels of vertical integration and growing monopolistic control over markets.

This turn of events, as we saw in Chapter 2, was based in part on the exploitation of resources and labour in colonial areas (Panikkar, 1959), which are called the *periphery* in Figure 7.1. After the Second World War, with the break-up of traditional political colonialism, a new means was needed to guarantee access to necessary production inputs. As well, capital intensive manufacturing techniques and high wage bills in the core area of the world economy brought with them a falling marginal return on investment, even under conditions of monopolistic competition. For these reasons, the *corporate sector* began to invest directly in overseas productive facilities—securing resources, lowering costs, and increasing profits. National corporations became multinational corporations, and gradually an important part of international trade came to be internalized within their corporate structure (Hymer, 1968, 1972a; Hymer and Resnick, 1971).

Up to this point the UDT argument is not different in kind from the analysis of national market formation presented in an earlier chapter. Here underdevelopment/dependency theory takes a different turn, however, for with the fall of imperial Europe the 'extended national peripheries' of the colonial powers became independent nation-states, led by Europeanized elites who were bent on creating their own industrial economies.

Now we can trace the various flows which link the three major components in Figure 7.1. At least at the outset, post-war expansion of the multinational corporation was seen as an obvious benefit to the industrial West. With Europe and its empire in shambles the *Pax Americana* filled the political vacuum for some thirty years, and aspirations for economic development in newly emerging countries fit nicely into the American scheme of world hegemony. Through the United Nations and various bilateral arrangements the United States would maintain global political dominance, and in turn industrial countries would secure needed resources and markets, multinational corporations would realize higher profits, and peripheral nations would be integrated into the new structure of the world capitalist economy (Nabudere, 1977).

The arrow at the top of Figure 7.1 shows this aspect of core/periphery relationships. According to UDT, the whole body of *political ideology* disseminated by the United States, its allies, and the various international organizations established after the War—the UN, IMF, the World Bank—were meant to legitimize and facilitate operation of the multinational corporate sector (Frank, 1971). Besides the general doctrines of free trade and market economy, a specialized body of economic theorizing was invented to promote these ends. This *modern theory of economic development*, in the hands of writers like Lewis (1955) and Rostow (1961), drew its ultimate inspiration from the classical political economics of Smith, Malthus, Ricardo, and Mill (Barber, 1967, pp. 107–115).

In a nutshell, for them, economic growth was based on specialization and the accumulation of capital. This could best be accomplished through a system of free trade in which different countries could reap whatever comparative advantages they might possess, and capital would be directed into the hands of those who would be most likely to reinvest it efficiently. In more specific terms, peripheral countries were advised to specialize in primary production and industrial activities which could take advantage of a low wage bill. The unequal distribution of revenue must be maintained to facilitate accumulation; in the event this meant allowing much of it to go to multinational corporations in the form of profit. To create an environment that would attract multinational participation, however, basic investments would have to be made in economic infrastructure and (to a lesser extent) in other forms of social overhead equipment. Because local savings and investment were deficient, this is where *economic aid* came into the picture. Core-country governments and international agencies would undertake to provide the shortfall in public investment in the form of loans and direct gifts in aid. This would be done for humanitarian reasons and in the name of enlightened self-interest.

Underdevelopment theory took a rather different view of these same relationships (Frank, 1971). In UDT terms, specialization in primary resource production meant in all probability becoming caught in a staple export trap, continuing to supply cheap inputs to Western manufacturers. Free trade meant allowing multinational corporations to dominate local markets and prolong this situation indefinitely, even destroying traditional local manufactures. A high share of national revenue or GNP to capital in such circumstances meant supporting international capital and a coopted local ruling class, while the majority of the population got little or nothing for their efforts. In fact, unavoidably, they would get (at least) relatively poorer, as agriculture turned to export activities and indigenous manufacturing succumbed to outside competition. Economic aid did more for the donor than the recipient, paying them interest on idle capital, providing markets for their surplus products, and tying peripheral countries into their sphere of political domination. To maintain such an obviously exploitative arrangement, once people in the periphery became aware of what was really going on, would require supporting increasingly repressive political regimes through the exercise of *military authority*, such as the numerous American incursions into various Latin American countries, Southeast Asia, and the Middle East.

Moving to the bottom of Figure 7.1, in the UDT model *commodity and factor flows* between core area and periphery are the most immediate material product of this new dependency arrangement. Unlike proponents of orthodox trade theory, however, underdevelopment theorists have treated these movements as primarily exploitative in character.

In the first place there is the question of unequal exchange (Amin, 1974; Emmanuel, 1972; Hymer and Resnick, 1971). The terms of trade between raw materials and semi-processed goods, which make up the bulk of exports from the periphery, and the capital goods and luxury items imported from the core area are notably 'unfair'.[1] This is the outcome of a world marketing system controlled by the multinationals. Primary products are purchased from a large number of producers by a relatively smaller number of buyers, giving the buyers a significant bargaining advantage. Because most of these commodities have relatively inelastic demand curves, market price is largely set by the quantity produced. Thus while it is in the interest of any one seller to produce as much as possible, competition between producers merely leads to a lower price for aggregate world output. With the resultant vagaries of market price this quickly becomes a zero-sum game, and with the exception of petroleum, where OPEC has managed to turn the game to its own advantage, for a while at least, historically there has been a marked tendency for the terms of trade in primary products to deteriorate.

Heavy manufactured goods—capital equipment and consumer durables coming from the core area—are in the opposite situation. Production is controlled by a small oligopoly of multinational firms which limit production and fix prices (Hymer and Rowthorn, 1970; Rowthorn and Hymer, 1971). This is a seller's market which operates on a cost-plus basis. As technical innovation, higher wage bills, and general inflation push costs up in the core area, these increases are passed along to consumers in the periphery, who are less and less able to pay for them out of declining revenues from primary products. Because demand from the periphery makes up only a small part of the world market for heavy industry, buyers have little or no leverage to counter such trends.

A second point which contributes to the exploitative nature of commodity and factor flows is the very low economic rent received by primary sector producers (Robinson, 1979). Because of the nature of the productive system, with marketing dominated by the multinationals and even productive facilities in the periphery frequently owned by expatriate investors, periphery residents and government get a very small share of the profits. In effect, they frequently end up almost giving their resources away. To aggravate the situation, since there is very little value added by resource extraction, the total pie is fairly small, encouraging stiff competition and conflict between different classes and government.

Finally there is the matter of opportunity costs (Robinson, 1979). Broadly defined, these include all of the local needs which might have been met if a country's primary resources were put to different uses, all the value added which might have been created by local processing, and the significantly increased capital formation which has been foregone. These are perhaps the most cruel features of the whole process. Resources are sold at give-away prices—many of them non-renewable—and people living in the periphery are left with only the memories of high potentials and bright expectations—a world of might-have-beens.

The central protagonist in the UDT scenario is the corporate sector, to which we must now turn for a more detailed look at the way in which it fits into the international economy. The arrows in Figure 7.1 connecting the corporate sector and periphery suggest some of the most important linkages. Direct foreign investment by industrial corporations was a novel feature of the post-Second World War economy. Continuing mercantile practices during the era of European domination had discouraged such activities by manufacturers. American firms pioneered the technique during European reconstruction and then went on to use it successfully in Third World countries.

The primal characteristic of direct foreign investment—the thing which makes it so very different from earlier patterns of economic integration—is that it tends to *internalize* the entire production and distribution process within the corporate control of individual firms (Hymer, 1972b).

Two of the most crucial factors which have been internalized are *technology and venture capital* (Michalet, 1975a, 1975b). For neo-classical growth theorists these are two of the most important services provided by the large modern corporation. Factors which peripheral countries are markedly short on are: (1) entrepreneurship and R&D and (2) working capital to pay start-up costs and provide a wage fund. For UDT writers they are also two of the most exploitative features of the contemporary world economy.

Technology is the mainspring of industry. This is equally true today for agriculture as for manufacturing. It is the dynamics of technical innovation and control within the multinational corporation that UDT uses as an indictment—struggles for market control force corporations to engage in research and development, with the aim of improving their entrepreneurial skills and designing more cost-effective fabricating techniques. Such expenditures, however, added to the subsequent costs of new capital equipment, bring with them larger-scale economies, better labour productivity, and higher wage bills. All of these things encourage, if not demand, that management seeks out cheaper sources of labour and expanded markets. This catch-22 situation, where profit maintenance entails automatic cost increases, is what has made investment in peripheral countries attractive to the multinationals.

When they arrive in a peripheral country they bring their technology with them. Know-how is one of the things the periphery is most anxious to acquire. But here is where the rub comes in, because knowledge is power and the multinationals only impart a discrete set of technical skills which are absolutely essential to the production process in question. This is part of the branch-plant syndrome: production units are increasingly scattered around the periphery, while the vital management and R&D skills remain in the core. It has also been argued that most of the skills transferred are 'consumption skills', allowing would-be customers in the periphery to buy and consume their products but providing little information about production processes themselves (Michalet, 1975a). Furthermore, highly capital-intensive technologies combined with rock-bottom wages may be good business for the multinational corporation, but they are hardly appropriate technologies for the development needs of labour rich/capital poor countries. Under these circumstances, 'technology transfer' helps provide the tools of exploitation and dependency.

Investment practices share many of the same characteristics. Venture capital is sought by peripheral elites to provide the productive facilities necessary for a modern industrial economy. To supplement meagre domestic capital formation, with the help of World Bank planners they devise elaborate incentive schemes to attract multinational investment, often providing much of the infrastructure and physical plant themselves through public spending and borrowing. When the multinational investor does arrive, their investment does little to stimulate the national economy; few jobs are created, few economic rents are collected, and few production linkages are established within the host economy. Most typically such plants act as export platforms to service corporate world markets. When geared toward the host-country market they tend to displace local production, putting local capitalists and traditional artisans out of business. Initial corporate investments are amortized very quickly, leaving companies free to move on to greener pastures if conditions warrant in as little as three to five years. Direct corporate investment is very lucrative business—for the corporate investors.

This brings us to the second arrow connecting peripheral areas to the corporate sector in Figure 7.1, flows of *resources and surplus value*. Physical resource extraction is a fairly straightforward proposition. When it takes place under the auspices of multinational capital raw materials and semi-processed goods enter the international market place as a component of corporate production. They may be sold as primary products or as intermediate inputs; the point is that they are withdrawn from other alternative uses and it is the multinational corporation which has captured most of their market value through selling them.

This broaches the very keystone of underdevelopment/dependency thinking, for in an exchange economy it is control of the surplus value of production that determines the outcome of the development process. There are apparently two sources of this surplus value: resource sales and the value added in manufacturing. As hinted above in the discussion of commodity and factor flows (see note 1), the 'value' of resources is very difficult to ascertain. None of the existing theories of economic value provide an analytical means of differentiating it from market price (Robinson, 1979). Although problematic, the best that can be done is to suggest perhaps that net profits on resource sales approximate the surplus value of the transaction.

Manufactured goods are an entirely different matter. Whatever the prevailing market price of an item, its surplus value is the sum realized above that required to remunerate the labour power used to produce it. Since workers in peripheral areas are forced to settle for very low wages in comparison with those in core countries, the difference between core and peripheral wage schedules represents the added surplus value reaped by the multinational corporation. All but some small portion of this total surplus value—economic rents accruing to government in the form of taxes or equity shares—is lost to the peripheral economy which has produced it—internalized into the multinational. The only other exception is retained earnings which are used for further investment within the country itself, and in this case host-economy workers are producing the means by which the multinational corporation avails itself of an ever-greater share of that economy's productive wealth and potential surplus value. Domestic capital formation is short-circuited and the corporation receives risk-free exploitation rights.

Closing the final link, then, the vast majority of surplus value realized in the peripheral economy is transferred back to the core in the form of *repatriated profits*. According to underdevelopment theorists, if capital formation is inadequate in most Third World settings it is largely because the core area is receiving the benefits of other people's work. While under contemporary conditions the immediate geographic beneficiaries of transnational reinvestment decisions are unclear (see the first part of Section 2 below), there can be no question that corporate profits and core-area white-collar employment benefit immensely from Third World resource exploitation and assembly operations. Some would argue that continued expansion and health of the capitalist economy are dependent upon them.

The bottom line of the UDT argument is that underdevelopment and dependency are both creations of the capitalist economic system. Underdevelopment is a process, the mirror image of development; while core areas develop the periphery is underdeveloped. Historically, one has seldom occurred without the other, and this sort of exploitation is so self-evident that its beneficiaries have frequently been forced to resort to political blackmail and military coercion.

Criticisms of the UDT model

As we shall see later in this chapter, UDT has been criticized by a number of writers, despite their general agreement with many of its fundamental assumptions. It is important to mention some of these objections here, however, in summary form, before we go on to examine UDT's application to peripheral regions in the industrialized countries. At heart the underdevelopment/ dependency model is a theoretical description of the *geographic transfer of value*, which concentrates on the macro-scale of the global economy. This is both the source of some of its most powerful analytic observations and its most evident shortcomings. While neo-classical models of economic growth placed the developing countries within a global marketplace, the real focus of most of their assumptions and propositions was the *national* economy. Underdevelopment theory has made an important contribution to allaying this idea of treating individual countries—especially those with small, 'open' economies—as meaningful economic accounting units. Instead, it has placed territorial development within a force field of powerful political, economic, and institutional influences. In an age of worldwide production and consumption, single countries, particularly new emerging states, exercise a questionable degree of political and economic sovereignty, as they are buffeted by the tidal currents of international power.

This is a telling perspective, but it also tends to turn a blind eye to other important aspects of Third World economies. UDT might be said to be far-sighted, paying insufficient attention to the domestic structures of dependency and underdevelopment. The parasitic nature of many Third World urban economies is recognized, and the class structure of Third World societies is linked to dominant international patterns. But neither of these fundamental contradictions in peripheral economies is given adequate treatment. While cities in peripheral areas are said to exploit their national hinterlands, this is attributed primarily to foreign domination and control. Little serious consideration is given to the dynamics of rural/urban relations in industrial society.

Similarly, UDT identifies the connections between local ruling classes and the dominant structures of international power. It correctly points out that peripheral elites tend to have Westernized values and aspirations, and are frequently on the payrolls of multinational corporations, or at least benefit indirectly from their activities. Underdevelopment theory goes little further than this, however. The class relations of an industrializing society are left largely unexplored, and the continuing poverty of the masses of Third World peoples is chalked off to the malevolence and greed of Westerners and the transnationals. Underdevelopment is a product of dependency. Local society and local rulers are substantially free from blame.

While this is an understandable reaction to the frequent powerlessness of peripheral nations, its comfortable xenophobia leads too easily to a conspiracy theory of underdevelopment. It may be a useful doctrine for maintaining political stability at home, but not necessarily an adequate framework for understanding social, political, and economic changes. The transformations of society which accompany the growth of a market economy are very complex, as we saw in Chapter 2. Contemporary 'dependent' development has many unique characteristics, qualities which differentiate it from earlier Western experience, but the production relations and social structures of peripheral countries still probably hold the keys to understanding this process.

What, in fact, are the real structural differences between 'dependent' development and other historical examples of development under industrial capitalism? Granting that the world situation is vastly different from 200 years ago, and granting that the multinationals represent a new actor on the economic scene, how are the broad mechanisms and outcomes of capitalist development different from the past? Capitalist development has always been uneven, with some people and some places benefitting from other people's labour and resources. The scale is obviously larger today and capital circuits take new and different forms, but is the development process not what it has always been— exploitation of surplus value, capital accumulation, and uneven development (Moore, 1979)?

Such observations are probably both useful and misleading, for like UDT itself they emphasize only certain aspects of the complex and contradictory workings of the contemporary global economy. The fundamental characteristic of this world system and thus 'dependent' development, as Wallerstein (1974a, 1974b, 1976a, 1976b) has pointed out, is its profound *interdependency*. While it would be mistaken to use such interconnectivity as a 'structural excuse' for legitimizing exploitative economic relationships (Brookfield, 1975), it would seem equally fallacious to look at only one part of the picture.

The unique events in the history of each country and region have, at any point in time, endowed that particular place with a specific economic productive capacity. This intertwined set of techniques, socially defined resources, and socioeconomic relations must be viewed in the context of the development of institutional structures within the broader world economy. It is this conjuncture or fit between local and global capacities and needs that, in general, determines the form of a locality's insertion within the dominant economic system and its probable development trajectory. Economic geography, as we shall see below, might best be conceived as a 'moment' of these relationships.

The increasingly dominant productive institution today, the multinational enterprise, is itself an outcome of development of the global economic system, and provides the nexus in which many of the economic relations between different classes living in different parts of the world are defined. Within this structure—because of the worldwide scope of production and consumption, the allocation and control of productive processes, and the eventual distribution of the surplus value created through production—development has many faces, few of them unambiguous. The picture painted by UDT theorists seems to describe one aspect of this reality. What can be said about the absolute increases in material living standards (e.g., money income, literacy, mortality rates) created by multinational activities in many Third World locations, especially the emergence of Wallerstein's 'semi-periphery'—such success stories in traditional neo-classical terms as Brazil, Mexico, South Korea, Malaysia, Taiwan, Singapore, Hong Kong, and the OPEC countries? Is this 'development' or 'underdevelopment' in UDT terminology? Like all basic ideological positions, underdevelopment/dependency theory does not lend itself to empirical verification or rejection at this level.

Similarly, what about the impact of recent institutional transformations on the economies of industrialized countries like France, Britain, and the United States? As we will see in the next section, divergences between the interests of transnational capital and the developed countries, and various regional economies within these countries, provide evidence of a rather more general contradiction between territorially organized political units and internationally organized economic institutions. Perhaps the most striking sign of this conflict is the increasing industrial 'restructuring' or 'deindustrialization' taking place in France, Britain, and the United States, as well as other traditional manufacturing areas (see, for example, Bluestone and Harrison, 1982).

2. UNDERDEVELOPMENT IN INDUSTRIALIZED COUNTRIES

The main lines of underdevelopment/dependency theory have been adapted by several authors to analyse conditions in the peripheral regions of France, Britain, and the United States, as well as other Western countries (e.g. Carney et al., 1975; Carter, 1974; Firn, 1975; Goodman, 1979; Holland, 1976a; Lafont, 1968; Mandle, 1978; Seers, Schaffer, and Kiljunen, 1978). While the UDT argument must be modified somewhat to reflect the nature of regional problems in industrialized societies, the fundamental themes fall clearly within the underdevelopment tradition.

Figure 7.2 suggests the different regional settings in which underdevelopment has occurred in industrialized countries. The first two rows, *resource exploitation* and *dependent development*, share much in common with the

Variety of underdevelopment	Type of region	Problem characteristics
Resource exploitation	Undeveloped resource frontier	Loss of economic rents, opportunity costs, and boomtownism
Dependent development	Newly industrializing area	Low wages, Extraction of surplus value
Reindustrialization	Metropolitan areas and 'enterprise zones'	Low wages, extraction of surplus value, and high social costs
Deindustrialization	Lagging manufacturing region	Loss of livelihood, creation of a reserve labour force
Internal colonialism	Minority culture area	Subdominance, historic exploitation and discrimination, dependent institutions

Figure 7.2   Underdevelopment in industrialized countries

concerns of writers in Third World countries. Resource exploitation takes place in regions like northern (off-shore) Scotland and the American West, where untapped reserves of raw materials and a sparse population make the area receptive to capital-intensive resource extraction by the multinationals. The primary problems here are those already mentioned in the Third World context: loss of economic rents, opportunity costs, and economic instability or boomtownism. Dependent development involves industrialization in peripheral regions with low labour costs and a thinly developed modern infrastructural base. These include areas like the French Midi and the American Sunbelt which have experienced rapid manufacturing growth of the branch-plant variety, attracted by a pool of unorganized labour (lacking alternative employment opportunities or a tradition of union activism) and easy access to metropolitan markets. 'Reindustrialization' in metropolitan core areas and enterprise zones, such as discussed at the end of the last chapter, attempts to promote high rates of accumulation by legally forcing low wages and ignoring the social costs of production. In essence, competing with Third World locations on their own terms by substantially recreating Third World conditions, insofar as possible (see Anderson, 1981; Goldsmith, 1981, 1982a, 1982b; Goldsmith and Jacobs, 1982; Hall, Castells, and Massey, 1982; Humberger, 1981; Mounts, 1982.

Capital versus the regions

Stuart Holland's work in England was probably the first to exert real influence among large numbers of regional scientists and planners. His writing concentrated on this latter question of dependent development, and the problem which most concerned Leo Klaassen in Chapter 5, lagging regions experiencing *deindustrialization*, characterized by a comparatively low or negative rate of economic growth. Holland proposed a model of increasing regional disparities based on imbalanced trade, increasing concentration, and variable rates of regional economic growth (Holland, 1971, 1976a, 1976b, 1979). The main argument was stated in terms of profit-maximizing (or least-cost) location theory, although later renditions (e.g. Holland, 1979) attempted to add a political/ideological component as well.[2] The prime mover in Holland's framework was the multinational corporation or *meso-sector*. After critiquing the equilibrium model of interregional growth, building on Myrdal, Holland substituted the factor flows suggested in Figure 7.3.

The immediate objective of regional policy had been to induce dependent development, encouraging the meso-sector to establish branch plants in lagging regions and hoping for the gradual accretion of beneficial economic multiplier effects. Although the French and American sunbelts might have experienced some semblance of this phenomenon, as part of a 'regional rotation' strategy by the meso-sector (Goodman, 1979; Lipietz, 1979a; Malizia, 1978), Holland argued that by and large peripheral Britain had missed out on even this type of development. Less developed regions (LDRs) in core

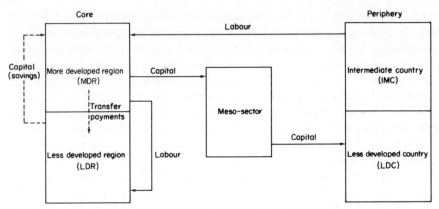

Figure 7.3   Holland's view of meso-selector control and global factor flaws

countries generated little surplus value for export to more developed regions (MDRs) because of comparative cost disadvantages *vis-à-vis* less developed countries (LDCs) in the world periphery. As shown in Figure 7.3, the main cash flow between LDRs and MDRs were savings (insurance and pension funds) and public transfer payments (see also Markusen and Fastrup, 1978; and Rifkin and Barber, 1978). Holland argued that by the mid-1970s there was a reverse migration of domestic labour back towards the less developed regions and that LDRs had probably come to experience a net capital inflow based on public sector transfers (e.g. welfare, unemployment, retirement benefits, and development incentives). The reasons for this had to do with the dynamics of international capital mobility and labour movements.

Even though integrated multinational production processes tend to be capital intensive, labour costs still typically make up the largest single production input. Thus, it is labour costs rather than transportation or fixed capital expenditures which determine corporate location decisions. Labour-intensive service activities and assembly operations which still take place in core-country MDRs, because of market orientation or legal constraints, use the cheapest available imported labour. Figure 7.3 shows this as international migration from intermediate countries (IMC) like Mexico, Algeria, and Puerto Rico—Wallerstein's 'semi-periphery'—to MDRs such as Southern California, the Paris Basin, and metropolitan New York. In turn, it is the competition of cheap foreign labour—legal and illegal—which is partially behind the movement of unskilled domestic labour back to the national periphery, added to the labour reserve in LDRs (see also Berry, 1980; Fuguitt, Voss, and Doherty, 1979; Morrill, 1979; Vining and Kontuly, 1978; Vining and Strauss, 1977; Zelinsky, 1977).

The most detrimental trend for LDR development is the outflow of core-country capital to LDCs, shown in Figure 7.3 as meso-sector investment—one of the most contradictory aspects of the workings of the contemporary capitalist economy, as suggested in the last section. Because of Third World labour costs, there are few LDR situations in which any combination of comparative advantage and regional incentive measures can begin to make up the differential causing widespread core-country deindustrialization. The dynamics of this process are suggested in the LDR/LDC cost schedules shown in Figure 7.4.

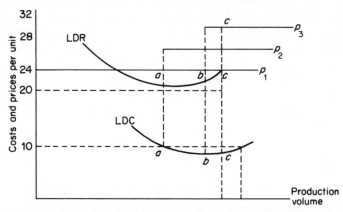

Figure 7.4   Comparative cost schedules for LDR and LDC locations

If p_1 is the MDC market price, based on the lowest-cost LDR plant location, a meso-sector firm investing in LDC production facilities gains a substantial comparative cost advantage across the whole range of its production schedule. At point a in Figure 7.4, LDC production costs are less than half those incurred by a LDR competitor. At point b, when marginal costs in the LDR location are beginning to rise and per unit profits fall, LDC production costs have fallen to something over one-third those required in the LDR location and the LDC meso-sector producer is still enjoying increasing scale economies. When the peripheral region core-country producer is forced out of the market, as point c, the LDC production unit is only beginning to experience diminishing returns to scale. While LDR profits at point b were 100 per cent. and had fallen to zero at point c, the LDC producer was reaping super-normal profits ranging around 200 per cent.

If the LDC meso-sector producer is an oligopolistic price-leader exercising a strong measure of market control, it can choose at any time to increase its market share and push the LDR producer out of business. On the other hand, a multinational corporation can also choose to fix the market price at p_2 or p_3, protecting profits at its own remaining MDC locations, fending off charges of monopoly practices by MDC governments (by allowing LDR producers to stay in the market, enjoying increased profits), and raising its own profit margins on LDC production to astronomic levels. As Holland points out (1979, p. 196), the meso-sector can inflate final price levels even further through transfer pricing or arbitrary inflation of imported components from foreign subsidiaries.

In such a situation deindustrialization is almost preordained, and MDC regional policies meant to help lagging regions are doomed to ineffectiveness. Even if MDC governments were to heavily subsidize both labour and fixed capital costs in LDRs, there would still be no cost advantage to attract the meso-sector, because in the extreme case, point c at price p_3 in Figure 7.4, there is a three to one cost differential and, more importantly, a four to one profit differential in favour of the LDC location. On top of this, LDC governments, with international assistance, frequently offer more attractive subsidies to capital than do MDC governments. The only remaining arguments for an LDR location, then, are non-economic ones: political appeasement, market control, and political stability. While these may prove to be the decision principle in particular instances, it is easy to see that the meso-sector is in an extremely favourable bargaining position *vis-à-vis* both LDCs and LDRs. In this scenario there are few winners except the meso-sector.

The internal colonialism hypothesis

Deindustrialization and doubtful prospects for even dependent development may be the economic grounds for LDR discontent, but there are other complicating factors as well. The most important of these is probably the internal colonialism hypothesis, shown as the bottom row in Figure 7.2 (turn back to page 122). This idea refers to the historic political and economic relations between a dominant national core area like South East England, the Paris Basin, or the American Northeast, and culturally dissimilar peripheral areas such as Britain's Celtic fringe, Corsica, Brittany, and Occitantia in France, or the US Old South. The emphasis here tends to be on military and political subdominance, especially longstanding grievances concerning exploitation, discrimination, and dependent institutions (Harvie, 1977; Hechter, 1975; Lafont, 1971; Persky, 1972, 1973). The internal colonialism hypothesis provides a format for consideration of a host of 'non-rational' factors frequently associated with regionalism or nationalism.

One of the most thoroughly developed and widely commented upon analyses in this mode is Michael Hechter's (1975) discussion of *Internal Colonialism: The Celtic Fringe in British National Development, 1536–1966*. Hechter, an American sociologist, attempted to understand the historical development of dominance/dependency relations within the British Isles, arguing that Great Britain could be characterized by a core/periphery structure representing, respectively, England and its surrounding Celtic fringe: Scotland, Ireland, and Wales. He went on to propose that relations between core and periphery had either been shaped by forces which could be described by a *diffusionist model of national development* or by a form of domestic imperialism in which the periphery was treated as an *internal colony*. He found the latter explanation more consistent with British experience.

While Hechter considered economic variables such as industrial structure, regional specialization and interregional trade, he felt that these were subsidiary in nature, primarily determined by *the form of political and cultural integration*. In defining the internal colonialism model he argued:

> ... the aggregate economic differences between core and periphery are casually linked to their cultural differences.
>
> In this description national development has less to do with automatic social structural or economic processes, and more with the exercise of control over government policies concerning the allocation of resources. Since increased contact between core and periphery does not tend to narrow the economic gap between the groups, national development will best be served by strengthening the political power of the peripheral group so that it may change the distribution of resources to its greater advantage. Ultimately this power must be based on political organization. One of the foundations upon which such organization might rest is, of course, cultural similarity, or the perception of a distinctive *ethnic identity* in the peripheral groups. The obstacle to national development suggested by the internal colonial model analogy, therefore, relates not to a failure of peripheral integration with the core but to a malintegration established on terms increasingly regarded as unjust and illegitimate.
>
> Thus the internal colonial model would appear to account for the persistence of backwardness in the midst of industrial society, as well as the apparent volatility of political integration. Further, by linking economic and occupational differences between groups to their cultural differences, this model has an additional advantage in that it suggests an explanation for the resiliency of peripheral culture (Hechter, 1975, pp. 33–34).

For Hechter, the central feature of internal colonialism is a *Cultural division of labour* enforced by dominant core-area political institutions which relegate the periphery to a position of economic dependency. Underdevelopment is a product of forced economic specialization and the lack of political sovereignty to protect peripheral markets and the continued development of well-rounded peripheral productive capabilities. Discussing the Celtic fringe at the time of political union with England Hechter concluded:

> The point here is that the state's responsibility to provide for economic welfare has continued in the *laissez-faire* and modern periods as well. The importance of states in economic development is obvious. They can coerce recalcitrant groups to accept government policies by using the threat of military sanctions. In addition to their ability to negotiate the terms of trade, states have the power to legislate against economic rigidities, to alter the rate of involuntary saving through taxation, and to mobilize collective sentiments to ends which they deem necessary. States alone can support large-scale activities which are not in themselves profitable, such as establishing transportation and education systems.
>
> Does it not therefore follow that the state is the greatest weapon in the arsenal of a society determined to struggle against economic dependency? This is not to say that political sovereignty alone is sufficient of the task. Almost everywhere the evidence points in the opposite direction. But in what territory has diversified development occurred without benefit of state protection? It is in this context that the loss of Celtic sovereignty must be considered. By denying these territories political independence, England made their increasing economic dependency inevitable (Hechter, 1975, pp. 91–92).

According to Hechter it is the loss of political sovereignty and the subsequent cultural division of labour among territorially defined ethnic groups which accounts for the economic disparities between regions. Peripheral populations, refused the possibility of economic diversification and barred from opportunities for assimilation and upward social mobility, become *collectivities* which are integrated on the basis of *status groups* rather than functionally defined economic classes. Culturally based group solidarity tends to overshadow all other considerations and reinforces existing cultural differences. In turn, economic underdevelopment and superimposed cultural identification lead to increasing *group consciousness* and undermine the limited legitimacy of existing political institutions. Eventually this juxtaposition of circumstances may bring about mass political movements in the periphery dedicated to achieving substantial political devolution or even autonomy. Hechter's empirical analyses of peripheral Britain were interpreted as lending support for this view, and he extrapolated that a similar phenomenon was the basis for contemporary regionalist movements in other European countries as well, such as France, and might even be used to account for minority group activism in the United States.[3]

Thus, the internal colonialism hypothesis throws a different light on the arguments of Holland and other economically oriented UDT theorists. While it accepts and supports their view that economic relations between core and periphery play an active role in peripheral underdevelopment, it asserts that the primary causal chain begins elsewhere. Adopting Hechter's line of reasoning, political subdominance leads to a cultural division of labour, resulting in dependent economic specialization and further cultural differentiation. Economic disparities, social rigidities, and status-group integration lead to peripheral political consciousness which can act as a stimulus to political activism and moves towards separatism. Obviously the economic and cultural variants of UDT theorizing are complementary rather than mutually exclusive. The appropriate interpretation in a given situation would depend on historical circumstances.

There are two important differences between them, however. First, the cultural/political view of underdevelopment provides an explicit explanation for peripheral regionalism and nationalism, which economic analysis alone is unable to deal with (see, for example, Carney, 1980; and Hudson, 1980). As Hechter argues, under conditions of evident status-group solidarity and regional militancy, rigid economic analyses seem more abstracted from reality than does the 'subjectively' founded internal colonialism notion. With continuing regionalist violence in both France and Britain and growing regionalist sentiments in the United States, regional theory can hardly afford to maintain its traditional silence on the issue of political activism. Second, and most importantly from a planning perspective, the internal colonialism approach suggests very different solutions to the underdevelopment problem than those proposed by economists like Holland. Holland, viewing the underdevelopment problem as a conflict between multinational capital and the nation-state, argues that the only feasible solution to regional problems is active intervention by the central government—meaning nationalization and public entrepreneurship. As we saw in an earlier quotation, Hechter recognizes the vital importance of state actions in economic development, but argues that regional underdevelopment can only be turned round by a substantial local say in the exercise of political power. In practice this probably means some important measure of regional devolution in political decision making and the attributes of sovereignty, as suggested by Proudhon and other planning precursors in Chapter 3.

3. RADICAL REGIONAL SCIENCE

Underdevelopment/dependency theory in its various forms has played an important role in calling into question the spatial equilibrium models which were widely employed by planners during most of the post-Second World War period. The initial resistance of neo-classical theorists has lost much of its credibility in face of widespread international acceptance of critical interpretations of regional problems. Even some introductory textbooks in fields related to regional planning are beginning to subscribe to elements of the underdevelopment argument (e.g. Cole, 1981; De Souza and Foust, 1979; Jumper, Bell, and Ralston, 1980). As critical analysts have attempted to expand the theoretical scope of their work, however, UDT has encountered a new source of criticism—the mainstream intellectual Left.

Underdevelopment theory began as a visceral reaction among Third World intellectuals to the apparent outcomes of capitalist development in the world periphery (e.g. Fanon, 1961; Rodney, 1972). From the late 1960s onwards incremental attempts were made to tone down its rhetorical vocabulary and strengthen its theoretical consistency. In this process UDT came to be significantly influenced by historical-materialist thinking, especially from French sources, and in so doing came more and more to adopt a Marxist view of the international economic system. In espousing Marxist interpretations of the extraction of surplus value, capital accumulation, class relations, and the laws of motion of capital UDT underwent a fundamental change in theoretical orientation, a change which has tended to undercut its own claims to validity. At the same time, UDT and the emergence of a Marxist school of urban sociology, first in France and then in Britain and the United States, have strongly underlined the need for a more explicit spatial dimension in Marxist economic analysis. The ferment which has resulted from the debate among Leftist scholars of varying inclinations has prompted the appearance of what might be called radical regional science or radical political economy. In this section I will outline some of the more important issues raised in this debate.

Uneven development versus underdevelopment

While the underdevelopment/dependency model began as an interpretation of empirical evidence, radical political economy of the Marxian school is founded upon a theoretical analysis of the capitalist economic system. Its fundamental concepts and analytic categories are drawn from Marxist economic theory, and, historically, have been used most frequently to provide a general framework for understanding the capitalist mode of production and its transformations. The application of these analytical tools to particular concrete situations is the goal of all serious Marxist scholarship, and their theoretical generality tends to lend such studies a very different perspective from that adopted by writers in the UDT tradition. In fact, it frequently puts them at cross-purposes with one another.

This becomes clear at the outset with formulation of the Marxist regional problematique. While UDT conceptualizes regional underdevelopment as a conscious outcome of core/periphery struggles and the designs of transnational capital, Marxists tend to see regional disparities as a *normal* product of capitalist economic development. In the anarchy of capitalist production, expansion takes place *unevenly* between various branches of industry, as well as between different geographic locations. Over time uneven rates of economic growth between different industries and regions will vary, but there is serious disagreement among radical regional scientists as to whether or not uneven development in its geographic dimension is *necessary* to the process of capitalist accumulation and the survival of capitalism (see Carney, 1980; Clark, 1980a; Mandel, 1976, 1978; Markusen, 1978b; Massey, 1978a, 1978b; Walker, 1978). For Marxists, however, *uneven development* is very different from the UDT concept of *underdevelopment*. The latter idea, it is argued, rests on the reified notion of 'places exploiting other places', while the former is ultimately founded in capitalist class relations and the rate of circulation of capital, commodities, and labour.

I will return to the Marxist explanation of regional inequalities below in discussing the State's role under monopoly capitalism, but this must be done in the context of a more detailed explanation of the historical-materialist debate about relations in the space economy. As with the percursors of regional planning discussed in Chapter 3, the classical writings of Marx and Engels recognized the critical role of contradictions between town and countryside within the capitalist mode of production (Marx, 1965, 1976; Marx and Engels, 1970). It was even suggested, as argued earlier here, that manipulation of spatial relationships was necessary for establishment of a class-based economy in which a small minority owned the means of production and the masses were transformed into a landless pool of wage labourers. In Marx's mature economic writings (i.e. *Capital*), however, the spatial dimension of capitalist economic relations was left unelaborated (Maarek, 1979; Mandel, 1968; Marx, 1909).

This set a pattern for later Marxist thinking which shared much in common with the 'anti-spatial' bias found in the work of many liberal political economists and sociologists. The economic and social relations of capitalist production came to be treated as if they existed in a *spaceless void*, and undue concern about geographic characteristics came to be labelled as a form of 'fetishism'. It was thought to be a liberal ruse used to mask the causal dynamics and primacy of class relations.[4] Marxist orthodoxy continues to assert this traditional position, and recently the felt need to reaffirm the class basis of economic disparities has been reinforced in response to a growing Leftist interest in the geography of exploitation (see Soja, 1980).

Besides the problematique of UDT and the underdevelopment/uneven development debate, there have been several other important contributions to radical spatial theorizing. Among the most relevant, perhaps, to the concerns of regional planners are: (1) the critical analysis of industrial location decisions; (2) the role of the capitalist State in uneven development; (3) a reawakening of interest in the contradictions between city and countryside; and (4) attempts to outline a critical theory of spatial organization and structure.

A radical theory of industrial location

Reconsideration of the industrial location problem in Europe seems to have come about largely through analyses of regional policy performance and the importance of transnational capital to industrial restructuring and the success of location and incentive programmes (e.g. see Cultiaux, 1975; Dunford, 1977, 1979; Massey, 1974, 1976, 1978c; Massey and Meegan, 1978, 1979; Michon-Savarit, 1975). In the United States changing regional industrial structures have prompted a similar critical response (e.g. Bluestone and Harrison, 1980a, 1980b, 1982; Clark, 1980b, 1981; Goodman, 1979; Markusen and Schoenberger, 1979; Rifkin and Barber, 1978). Out of the morass which characterized liberal theories of industrial location, by the early 1970s (Webber, 1972) radical analysts had identified a new pattern and rationale in corporate decision making.

Through the first decade or so after the Second World War industrial development tended to follow the broad geographic patterns set in the latter part of the nineteenth century. Economic growth in France, Britain, and the United States was led by the expansion of secondary industry, and manufacturing, apparently attracted by various urban scale economies, became increasingly concentrated in the traditional urban core areas (see Chapter 2). While there was an increasing flight of economic activities to the major metropolitan corridors, regional specialization was still dependent in most places upon the forward and backward linkages that developed around key local industries serving national and international markets. All the various phases of production—management, assembly operations, and divers support services—tended to take place in the original home region. So, for example, Detroit and Coventry were known for automobiles, Lancashire and Lyons for textiles, and so forth.

With the changing international structure of capital all this began to undergo profound changes by the 1960s. First it became clear that some of the older manufacturing regions, such as Nord in France, New England in the United States, and the British North East—areas which had long been in decline—were now irretrievably lost as centres of industry. Regional policy had contributed little to shoring up their collapsing economies. Although this could be attributed to the continuing metropolitan exodus there were also other conflicting signs. While the large cities apparently continued to grow, manufacturing plants began almost unaccountably to spring up in the midst of farm fields and declining rural service centres. New regional growth areas were recognized in the southern United States and France, and yet it was not until the late 1970s that it could be said for sure that growth rates in the metropolis had begun to taper off (Moriarty, 1980).

This was the quandary which faced location theorists. None of the traditional formulas seemed to work any longer: neither existing infrastructure, skilled labour pools, urban economies, nor market orientation seemed to explain what was going on, and few of the new manufacturing plants could be said to be resource oriented. Most of them were assembly operations—mere branch plants (Lonsdale and Seyler, 1979). Some liberal economists, planners, and geographers wrote of the growing uncertainty in locational decision making (Webber, 1972). Some came to the conclusion that the whole decision process was frankly irrational or that the friction of distance had ceased to count for much (Webber, 1964). Others adopted a more knowing attitude, asserting that upon closer inspection it could be shown that all the new trends had been predicted in the existing literature (Alonso, 1978). This made them no less confusing, however.

Three Marxist scholars, Christian Palloix (1975a, 1975b, 1975c, 1977) and Alain Lipietz (1977) working in France and Doreen Massey in the United Kingdom, managed to throw new light on the problem. Drawing on ideas about the internationalization of capital and the liberal literature on 'production stages' they identified a new interregional and international division of labour under monopoly capitalism.[5] In some respects their arguments are not dissimilar from the UDT analysis of dependent development, but they are cast within a much more general view of the motion of capital.

For Palloix, the apparent tendency towards geographic equalization of the conditions of production is offset by a new differentiation of productive activities from place to place. The major force behind these changes in industrial location patterns is not the institutional decision-making capacity of the multinational corporation, but rather the necessary contemporary modes of the *mise en valeur* or self-expansion of capital. This means the realization of value by the motion of capital through its three circuits (i.e. commodity capital, finance capital, and productive capital) and the expansion of the value of capital through the action of labour upon it. According to Palloix, historically, this self-expansion of capital has been concentrated in different capital circuits. First, in the era of 'free enterprise', it was centred around the international exchange of commodities, when value was added to and realized through the exchange of goods. Later it was money-capital which was internationalized, with the circulation of financial assets acting as the main route to expansion. Today, the redeployment of capital is taking place in the sphere of production itself, and international standards of value are replacing national value. The multinational corporation with its far-flung production facilities is the outcome rather than the cause of these changes. This is a process of simultaneous centralization and specialization which is common to all capitalist development.

Lipietz and Massey have attempted to work out the implications of this process for the spatial division of labour, arguing that geographic differentiation in production is largely dependent on a region's capacity for capital accumulation. The redeployment of productive activities under centralized monopoly control has resulted in a shift from regional specialization by the *industrial sector* to specialization by the *production stage*. Whereas traditionally all the phases of production of a given commodity were localized around a particular production point, today the various stages of production are located with reference to maximizing the geographic incidence of the accumulation of capital.

In concrete terms, R&D and management functions, which require the highest skill inputs but produce the largest expansion of capital value having the highest organic composition, occur in locations with the best accessibility to skilled labour and money-capital. These tend to be metropolitan areas in core countries of the world economy like South East England and the Paris Basin. Heavy industry and industrial engineering continue to make use of productive capital and organized labour in traditional manufacturing regions, although these areas, such as the American manufacturing belt, are today in marked decline. Wage goods, requiring only standardized assembly techniques and semi-skilled labour, are produced in peripheral regions, either in the rural fringe of industrialized countries or the Third World. The pattern which emerges at the regional level is tertiarization of the metropolis, deindustrialization of the traditional 'black belts', and the seeding of the countryside with isolated assembly plants (Carney, Hudson, and Lewis, 1980; Lipietz, 1979a).

Uneven development and the capitalist state

This new geography of uneven development, which accompanies the redeployment of capital, is actively supported by the class-based capitalist State. Ultimately the continued existence of the State is dependent upon the efficient extraction of surplus value and its realization through the circulation of capital. So despite the social and political problems which follow in the wake of industrial rationalization, the self-expansion of capital under the law of international value requires that the State intervene on behalf of internationalization (Carney, Hudson, and Lewis, 1980). This pits the State apparatus against the immediate interests of some regional segments of capital (Dulong, 1976a, 1976b, 1978), helping to forge the reactionary alliances between regional labour and capital common today in France, Britain, and the United States. As we saw in the last chapter, regional policy becomes a tool for attempting to legitimate the loss of local jobs and investment opportunities, while at the same time promoting industrial restructuring and the international circulation of capital (Regional Social Theory Workshop, 1978). The State thus becomes a central actor in creating contemporary patterns of uneven development, but in so doing the central contradictions of monopoly capitalism are shifted squarely into the political realm (Carney, 1980). The economic crisis—including regional economic problems—results in a multidimensional crisis of the State (Habermas, 1973; O'Connor, 1973; Poulantzas, 1976b). The struggle for regional economic development is transformed into a political struggle at the national and international levels (Regional Social Theory Workshop, 1978).

Analysis of the State's role in the crisis of the 1970s and '80s has become a central focus of debate among critical regional scientists.[6] At a general level two fundamental questions have been raised: (1) What is the relative autonomy of State activity? and (2) What are the inherent contradictions involved in State intervention? Figure 7.5 summarizes the domains of State activity and intervention. While it is outside the scope of this discussion to analyse each in detail, we must briefly consider the relationships which bear most directly on regional development and planning.

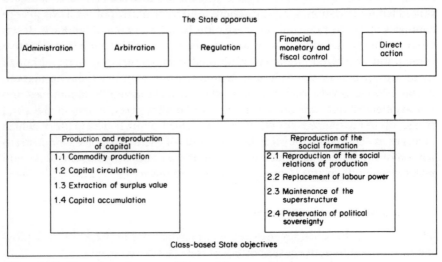

Figure 7.5   Domains of state activity and intervention

The State apparatus includes five major spheres of intervention, ordered from left to right in Figure 7.5 according to their relative degree of activism. *Administration* probably represents the closest approximation to the eighteenth century liberal ideal of neutral government—what Proudhon and Marx referred to a hundred years later (in the context of a classless society) as the 'mere administration of things'. *Arbitration* and *regulation* are more active applications of the police power, settling disputes between different groups in society and attempting to control their activities so as to prevent open conflicts from arising. *Financial, monetary and fiscal control* represents government activism in consumer markets, the money supply and capital circuits; i.e. direct and indirect action in the economy through decisions by the public treasury.

Direct action implies positive public sector activity, usually either economic or military. At the margin this latter domain of direct State action probably overlaps all the other four, indicating more the degree of State initiative involved than the kind of activity.

These various instruments of governmental power are used to further two general class-based objectives: (1) *production and reproduction of capital* and (2) *reproduction of the social formation*. The first of these entails encouraging in every way possible the *increased production of commodities* (1.1), the more rapid *circulation of capital* (1.2), more efficient *extraction of surplus value* (1.3), and the further *accumulation of capital* (1.4). On the other hand, the latter objective, reproduction of the social formation, means guaranteeing the continued *reproduction of the relations of production* (2.1), assuring the *replacement of labour power* (2.2), providing for *maintenance of the superstructure* (2.3), and ultimately the *preservation of political sovereignty* (2.4). Under the logic of capitalism, maintenance of the social formation is *essential* for, but *auxiliary* to, commodity production and the accumulation of capital. This relationship introduces a fundamental contradiction in the activities of the capitalist State, and in itself requires that some degree of autonomy be exercised by the State in attempting to carry out its mutually antagonistic objectives. It is in this autonomous action space that the battles of social reform take place and the seesaw course of bourgeoise policy, as we saw in the last chapter, alternates between liberalism and conservatism, trying to come to grips with an impossible combination of tasks.[7] In the context of regional development and planning the most striking contradictions in State activities today are probably brought about by the confrontation between footloose multinational capital and the reproduction needs of the nation-state. Regional groupings of labour power, capital, and sub-national governments find themselves impacted in a multiplicity of ways, creating the congeries of incompatible needs and demands which characterize the contemporary crisis.

The contradictions between city and countryside

Redeployment of capital and the changing geographic patterns of economic production have also reawakened critical interest in the relations between a city and its surrounding region—what the precursors of regional planning called the contradictions between city and countryside (Friedmann, 1979; Harvey, 1973; Merrington, 1975; Mumy, 1978–1979; Woodruff, 1980).

From the middle decades of the nineteenth century the industrial city had been in a continuing state of metamorphosis. The concentration of industrial activities within a medieval shell in France and Great Britain led to an eventual explosion of urban-like development across the countryside. This happened even more readily in the open cities of North America. Rural social relations, which had been destroyed by the penetration of wage labour into the countryside, were now a distant memory, as factories began once again to make their way into formerly rural areas. As we saw in Chapter 3, Patrick Geddes coined the name 'conurbation' to describe this maze of built-up land occupied by traditionally urban economic functions, and fifty years later Jean Gottman applied the term 'megalopolis' to the urbanized eastern seaboard of the United States (Gottman, 1959). Agriculture too had become merely another branch of industry; farms were simply open-air factories, operated from comfortable suburban homes or faraway corporate headquarters. All differentiation between the urban and the rural seemed to have come to an end, and regional scientists began to treat the humanized landscape as an isotropic plane characterized by a greater or lesser degree of urban development (e.g. Berry, 1973; Friedmann and Miller, 1965).

The crisis of the cities and deindustrialization of metropolitan life, however, have forced a reconsideration of rural/urban relations. As with macro-scale models of industrial location, traditional theories of the urban space-economy and city/hinterland relations seem in need of significant revision. At the level of the city itself this has given rise to a new school of critical urban scholarship which is beyond the scope of the present study to review (see, for example, Castells, 1972; Dear and Scott, 1980; Harloe, 1976; Harvey, 1973; Lefebvre, 1968, 1970, 1972; Pickvance, 1976). The questions of intraregional structure and the changing nature of rural/urban relations have received less attention, but there are indications of new interpretations here as well.

Historically, the geographic division of labour between a city and its hinterland may have been the first structural specialization of economic production (Merrington, 1975). Under capitalism it formed one of the bases for commodity production and exchange, and thus helped provide the necessary mechanism for the realization of surplus value. Specialization of economic pursuits between town and countryside was founded more on political suzerainty, economic dominance, and exploitation than any calculation of comparative advantage, as suggested by Adam Smith (Mumy, 1978–1979). As a feature of the capitalist economic system, rural/urban relations will necessarily undergo change with the transformations of the capitalist mode of production. The 'planned deurbanization of the metropolis, dissolving the city into the "urban region" ', according to Merrington (1975, p. 88), is one such transformation. He goes on to argue, citing Bookchin (1974), that: 'The mobility of mature social capital presupposes this capacity to reconstitute the town–country division on an ever-renewed basis. . . .'

The contemporary transformation of rural/urban contradictions should thus be conceptualized as part of the 'globalization' of the class conflict (Friedmann, 1979). New forms of rural specialization are no longer necessarily tied to their immediate metropolitan hinterland, but are part of the new global division of labour (Friedmann and Wolff, 1982). Contradictions are expressed through the lack of parity between agricultural and industrial prices (unequal exchange) (Merrington, 1975) and the dependent forms of rural industrialization (by production stage). Rural/urban relations, as in the beginnings of capitalist industrialization, reflect the more general core/periphery structures of dependence and exploitation.

Towards a Marxist theory of space economy

It is from the recognition of the essential unity of 'spatial political economy' and its integral role in economic production that Soja (1980) calls for a new dialectical materialism that is 'simultaneously historical and spatial'. Recalling the debate over *underdevelopment versus uneven development*, trends in Marxist urban analysis and the controversy on the Left about spatial fetishism, Soja proposes that spatial structure should be seen as part of a sociospatial dialectic.

Social or created space, like social relations, is an essential characteristic of economic production. Although not elaborated in most of the Marxist literature, spatial structure is a 'horizontal' homologue of 'vertical' class structure, and both are founded in the class relations of production. According to this line of reasoning:

> ... the vertical and horizontal expressions of relations of production under capitalism (i.e., relations of class) are, at the same time, homologous, in the sense of originating in the same set of generative structures (e.g., the relation between labour and capital); and dialectically linked, in that each shapes and is simultaneously shaped by the other in a complex interrelationship and at different historical conjunctures. There is no permanent and rigid dominance of one over the other in all concrete historical and geographical circumstances. Indeed, the historical development of the dialectic between social and spatial structures—the interplay between the social and territorial division of labour—should be a central issue of concrete Marxist analysis (Soja, 1980, pp. 224–225).

This perspective provides a means of summarizing the various radical analyses of regional problems presented in this chapter. Changing patterns of regional development and underdevelopment are a spatial manifestation of transformations in the capitalist mode of production. Unequal exchange, labour and resource exploitation, and the geographic transfer of value are all components of this process. They are shaped by and help to shape the class struggle between labour and capital. Contemporary changes in the location of economic activities represent a necessary redeployment of capital—an attempt to increase the rate of capital growth and circulation. The multinational corporation has become the main institutional agent in this internationalization of production, aided and abetted by the class-based capitalist State through such measures as regional planning. Regional problems have thus become primarily political problems, threatening the legitimacy of the State itself, and the class struggle at this point in history, in one dimension, becomes a regionalist struggle.[8]

4. CONCLUSIONS

As we saw in Chapter 6, governments in France, Britain, and the United States have retreated from the regional policies of the 1960s and early '70s. The inability of regional science-based planning to find a solution to continuing regional disparities has led to a new, radical reformulation of the problem. This conflict model of regional development is lent a measure of credibility by the increasingly transparent relationship of geographic inequalities to the workings of the international economy and the political activism of various regionalist movements in all three countries. Planners find themselves confronted for the first time in several generations with a situation in which *policies* become an obvious synonym for politics and strategies a synonym for *struggle*.

But what is the role of regional planning in such a politicized environment? Is there an approach to development for regions in crisis? Can there be a relevant political strategy for regional development and planning? These are the questions we shall address in the last part of this study. There are presently few generally accepted answers, but I believe the broad outlines of an approach to development in different regional settings can be suggested. While concentrating on contemporary issues, in many ways this perspective recalls views held by the precursors of regional planning, such as Proudhon and Kropotkin, discussed in earlier chapters.

NOTES

1. Joan Robinson gives an especially revealing discussion of the problem of determining the exchange value of primary products, around which this whole question of fairness revolves (see Robinson, 1979, pp. 63–66).
2. I would like to thank Stuart Holland for introducing me to some of his more recent ideas about politics, ideology, media control, and the role of the State in capital/labour relations.
3. When Hechter wrote in the early mid-1970s renewed regionalist movements had yet to attract serious attention in the United States. It is problematic whether he would have attempted to use the same basic analogy in the American context, but some American writers on the Left have gone to great pains to discredit such an interpretation (e.g. see Markusen, 1978b).
4. Certainly planning precursors such as Le Play and Geddes were guilty in varying degrees of such charges, and regional planning of the comprehensive river basin school during the 1920s and '30s contained a strong measure of environmental determinism which downplayed the importance of class relations. Urban and regional planning in the 1960s and '70s was almost entirely devoid of any serious social-class analysis (see Bookchin, 1973; Castells, 1973; Harvey, 1973).
5. While recent critical analyses of industrial location attempt to place locational decision making within a Marxist framework, frequently they fail to identify the historical sources of the two most central features of their model: (1) the contemporary spatial division of labour and (2) its subsequent 'stages of development'. The first work in these areas was done by liberal researchers at the Harvard Business School, working on the idea of 'the industrial product cycle' (e.g. Vernon, 1959, 1966), and Gunnar Törnqvist and his colleagues at Lund University in Sweden who did research into the behaviour of 'contact systems' (Thorngren, 1970; Törnqvist, 1970). The importance of this omission is that these ideas, taken in tandem with Marxist theory by writers like Massey, can and were in fact, originally founded in entirely different intellectual milieux, and lend themselves to various sorts of syntheses. For further criticism of these formulations see Weaver (1981b).
6. A general presentation of various Marxist positions may be drawn from Bobbio (1978), Domhaff (1979), Habermas (1973), Jessop (1977), Holloway and Picciotto (1978), Laclan (1975), Lebas (1979), Offe (1975), Offe and Ronge (1975), and

Poulantzas (1976a, b). Special attention is given to the domain of State activity labelled *financial, monetary and fiscal control* in Figure 7.5 by O'Connor (1973), Gough (1975), Markusen and Fastrup (1978), and Miller (1978). Aglietta (1979) presents a theory of capitalist regulation. Theoretical discussions and empirical analyses of regional and local aspects of State activity and intervention are given in Bleitrach (1977), Bleitrach and Chenu (1975), Carney *et al.* (1975), Carney, Hudson, and Lewis (1980), Clark (1980a), Davies (1974), Dear (1980), Dulong (1976a, 1976b, 1978), Dunford, Geddes, and Perrons (1980), Ewen (1978), Geddes (1979), Goldstein and Rosenberry (1978), Hudson (1980), Jensen (1980), Levy (1979), Lojkine (1977), Markusen (1978b), Morgan (1979, 1980), Pickvance (1977), and Regional Social Theory Workshop (1978).

7. The involved contradictions and immediate complexity of State activities and interventions disallow overly simplistic theoretical associations between the State and any particular segment of capital or mechanism of accumulation; this is complicated manyfold, as we will see in more detail in the next two sections, when the element of space is taken into consideration (Asheim, 1979). Even before adding the spatial dimension, however, constructing a 5×8 matrix from the generalized activities and objectives shown in Figure 7.5 would already create a complex action space involving forty different activity/objective combinations. The possibilities of contradictions and conflict abound even at this abstract level; the concrete reality is significantly more complex.

8. Soja (1980, p. 224) argues:

> ... the transformation of capitalism can occur only through the combination and articulation of a horizontal (periphery vs. center) and vertical (working class vs. bourgeoisie class) struggle, by transformation on both the social and spatial planes.
>
> The two forms of class struggle can be made to appear in conflict, especially with the manipulation of territorial identities under bourgeois nationalism, regionalism, and localism. But when territorial consciousness is based on the exploitative nature of capitalist relations of production and reproduction, and not on parochialism and emotional attachment to place, it is class consciousness.

CHAPTER 8

Regions, Wealth and Economic Location

The central thesis of radical regional science as discussed in the last chapter is that the capitalist economic system *necessarily* creates uneven development. In terms of social classes, industrial sectors and geographic space the structural relations of production and consumption are inherently unequal. These relationships change over time in response to transformations of the capitalist mode of production, but inequality in some form is an integral part of the system,[1] resting ultimately on production relations and the social and geographic division of labour. This is essentially the argument of the precursors of regional planning discussed in earlier chapters—updated and presented in the vocabulary and conceptual framework of contemporary political economy.

In the late twentieth century the world economy has undergone profound institutional and locational changes. Multinational corporations control a significant share of global productive activities, frequently exercising near-monopoly power in allocating investment, determining factor mixes, setting production schedules, fixing prices, and extracting profits. Individual countries, even in the historic global core area, appear relatively impotent in their attempts to safeguard domestic production and jobs, creating a perilous balancing act for the capitalist state. Sub-national regions in France, Britain, and the United States find themselves caught between the parrying of government and corporate interests. Outworn industrial areas are abandoned. Cheap labour is exploited in widely scattered factories. Resource regions are lured into the cycle of short-term boom and bust. More and more metropolitan areas become drones—losing their endogenous manufacturing capacity. Most recently, some like Los Angeles are being 'Latin Americanized', meaning *reindustrialization* on the Third World model (see *Business Week*, 1980).

Theoretical interpretations of these phenomena vary, not necessarily following broader ideological lines. While few would argue that regional deindustrialization has many positive attributes, the dynamics of branch plant development, resource extraction, and reindustrialization are much more con-

Portions of Chapter 8, Sections 1 and 2, have appeared in earlier form in *London Papers in Regional Science*, Vol. 11, 184–202.

tentious; the earlier regional convergence debate discussed in Chapter 5 has reappeared (see Berry, 1980; Clark, 1980a; Dunford, Geddes, and Perrons, 1980; Markusen, 1978b).[2] However, unless we are to fall back on Williamson's (1965) and Klaassen's (1965) equation of regional income with regional development—or other equally descriptive consumption measures—the regional development problem must be formulated in rather different terms. The relevant questions for understanding *regional crisis formation*, as we saw in the last chapter, have to do with the structure and processes of economic production.

From a regional development perspective there are a number of essential issues which must be confronted. In this chapter I will focus on two questions which I believe to be of paramount concern in any reconsideration of regional theory. First I will ask: What is the relationship between the production of economic value, regional specialization, and exports? And second, what are the contemporary economic constraints on locational decision making? I will argue that the answers to these questions define recognizable limits to economic determinism in regional affairs and identify important leverage points for a political strategy of regional development and planning—a situation vastly different from that facing early regional planners during the 1920s and '30s, when—despite the Great Depression and Second World War—urban-industrialization, rural decay, and core-area dominance were rolling ahead with the momentum of Destiny. The political economy of the late twentieth century may well provide precisely the conditions necessary for the redevelopment of more sound, self-reliant regional communities in France, Britain, and the United States, as envisioned by Proudhon and Kropotkin a century earlier.

1. THE SOURCES OF ECONOMIC WEALTH OR VALUE[3]

First we must look at the question of regional wealth. Figure 8.1 portrays the sources of economic wealth or value in a space economy. *Labour power, physical resources*, and *capital*, of course, represent the primary factors of economic production; use value and exchange value are the outcomes of that process. A region may employ its own labour (LP), resources (PR), and capital (K), or these can be brought in from outside the area (i.e. LP', PR', and K'). Use value (UV), meaning goods and services created for the benefit of their producers, is normally site-specific, typically being utilized *in situ*. Likewise, potential exchange value (PEV), commodities produced for sale in the market economy, is initially an attribute of economic activities which occur at specific geographic locations. It is only when its value is realized through sale, i.e. realized exchange value, either within the region of origin itself (REV) or somewhere else (REV'), that it enters the capital circuit and can be enumerated.

Although these fundamental concepts are quite familiar, as should be apparent from the discussion in earlier chapters, their implications within the regional setting have seldom really been explored. Theoretically fixed

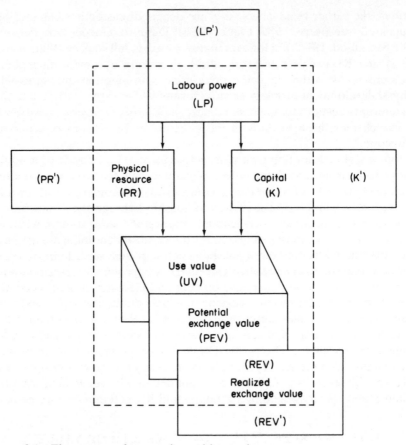

Figure 8.1 The sources of economic wealth or value

boundaries have limited the view of regional wealth creation to a narrow range of alternatives. This reductionism has then been accorded scientific legitimacy and political sanction, built into the institutional structure of government, and made the linchpin of economic policy—as well as the target of radical criticism. We are hard put to think outside its terms of reference. I will argue here, however, that such a particularistic viewpoint has identifiable historic origins and serves explicit ideological purposes (Myrdal, 1969). Furthermore, to achieve substantial change under contemporary circumstances it may be necessary to adopt an entirely different perspective, based on broader definitions of economic value creation in a regional economy. This will require that we reconsider in capsule form some of the key ideas covered in the preceding chapters.

Classical economists in the eighteenth and nineteenth centuries were anxious to identify the origins of wealth in a national economy. Their search began with analysis of the mechanisms which determine the 'natural' price of commodities and carried them to formulation of the labour theory of value (Marx, 1967; Ricardo, 1953; Smith, 1970). In this context it was only a short step to identifying the general benefits of encouraging comparative advantage through the division of labour and economic specialization (Ricardo, 1953). In fact, as writers in the classical tradition from Smith to Marx pointed out, primitive accumulation and the foundations of capitalist production were dependent on specialized labour and exchange (Barber, 1967; Deane, 1978; Heilbroner, 1953). Even more fundamental than the question of economic efficiency was the imperative to develop a wider spectrum of needs which must be met through trade. Only then could true commodity relations be established and only then could significant amounts of surplus value be realized, creating in middleman profits the first free form of capital (Bookchin, 1974). The political creation of wage labour through such devices as the Enclosure Acts and the Poor Law in England allowed this process to be pushed back a step further, extracting potential surplus value at the point of production itself and laying the basis for industrial capitalism (Hobsbawm, 1962) (see Chapter 2).

While it was recognized that the first widespread division of labour probably took place between the town and countryside (Merrington, 1975), the local geography of production and exchange was a matter of minor concern. As we saw in Chapter 2, both mercantilism and industrial capitalism in Europe were based on national class alliances between the bourgeoisie and aspiring absolutist monarchy. The national market provided an economic basis for the modern nation-state as well as the geographic basis for capitalist trade and accumulation. The Liberal battle for Free Trade throughout the nineteenth century was an attempt by the predominant powers to extend potential markets, and thus capital circuits, over ever wider geographic areas. Protectionism on the part of newly industrializing countries like the United States and Germany was an extremely telling precaution. Although capital might be brought in from outside, the realization of surplus value added by manufacture had to be limited as far as possible to entrepreneurs within the national economy if local accumulation was to occur.

We can trace historical developments and the arguments of the theorists with the aid of Figure 8.1. *Realized exchange value* (REV), in the lower right-hand corner of the diagram, was recognized to be the source of *capitalist accumulation* (K). Profit gained through selling commodities in the market was the major circuit of such accumulation under merchant capitalism; the amount accumulated was a determination of market size and selling price. Theorists traced the question of price to a commodity's *potential exchange value* (PEV) and ultimately to the *labour power* (LP) necesssary for its production. Market size rested upon the geographic area opened to trade, typically an artifact of political boundaries, represented in Figure 8.1 by the broken line, and the sphere of economic life which could be directed away from the use value (UV) and simple commodity production and refocused around market transactions (Bookchin, 1974). In Europe both these matters were in part, if not predominantly, political issues, resting on the breakdown of guild monopolies and price fixing, the establishment of larger polities, and the replacement of feudal obligations with money payments.[4] The economist's doctrine of Free Trade was an attempt to push market boundaries beyond the frontiers of the nation-state (REV'), a political device with profound economic implications that could only work themselves out with industrializatic n.

Merchant capitalism had been a mechanism for organizing, expanding, and exploiting the *distribution* side of the economy; the genius of industrial capitalism was that it cóntrolled and developed *production* as well. The factors of production themselves became commodities which could be controlled by capital and organized systematically for the creation of potential exchange value (PEV). This new efficiency, aided by political demands of the rising middle class, and eventually the policies of liberal governments, allowed capitalist manufacturing to displace most use value (UV) and simple commodity production. The possibilities for creating and realizing surplus value were multiplied manyfold, and reinvestment of surplus value in productive capital established new capital circuits which endowed the entire process with a revolutionary dynamism. As manufacturing grew and destroyed the guild system of production it was able to move into the towns, through its expansive energies and various legal contrivances, creating the vast proletarian conurbations of the nineteenth and early twentieth centuries (Bookchin, 1974; Merrington, 1975).

Within the nation-state productive forces became highly mobile, concentrating in established industrial centres: people, resources, and capital (LP', PR', K') could be brought together from various regions according to the capitalists' ability to realize exchange value (REV, REV') and pay for them. A monopoly over techniques, a headstart in accumulation, and direct access to political power gave ascendent metropolitan entrepreneurs an overwhelming advantage. Thinking of Figure 8.1 within the context of sub-national regions, capital from the metropolis (K') was able to buy up local productive forces (LP, PR, K) and redirect them towards its own ends (PEV), selling some of the commodities inside the region itself (REV) and some outside the area (REV'). Production factors could also be purchased and then transformed and sold in other regions. In either case, capital circuits in both Europe and North America were redirected to centre on the metropolis and a kind of regional specialization was established, if only by attrition in peripheral areas. Based to some extent on the peculiarities of history and physical geography, some places—the metropolis—built up an increasing fund of capital and productive capabilities, while others found themselves with less and less wealth (UV, REV, REV') and fewer and fewer things to do (UV, PEV).

Moving back to the national scale, the concept of Free Trade advocated the creation of these same circumstances between countries. Commodity markets would be expanded (REV'), as would factor markets (LP', PR', K'). Of course, countries exporting resources (PR') and those exporting capital and labour (K', LP') would typically be different. And it would be the countries which controlled capital circuits and determined the nature of production which would be the most likely to benefit. Without this power any increase of potential exchange value brought in from outside the country and realized in the local economy would only mean fewer possibilities for local production and accumulation.[5] Countries which came into control of modern productive forces experienced economic expansion and their middle classes flourished; countries that supplied mainly raw materials (PR) and created little indigenous potential exchange value (PEV) became relatively poorer and increasingly dependent.

Despite the repeated use of protectionism to foster national industrialization wherever circumstances allowed, the traditional liberal doctrine of *laissez-faire*, under the mantle of Free Trade theory, has remained the centrepiece of economic thinking up to the present day (see, for example, Heller, 1968). Emanating first primarily from Britain and then the United States, Free Trade has been enforced whenever possible, on others, as the basis of international relations. Exchange value realized through exports (REV') has become almost the only acknowledged source of wealth, for reasons we will explore more thoroughly in Section 3 below. Concepts such as comparative advantage and the benefits of geographic specialization became handmaidens to this idea. Their origins in the writings of the classical theorists and modern ideological purpose—with multinational capital expanding exchange value at every stage of production—seem clear. We must now turn to regional economics to appreciate their place in regional development and planning.

2. ECONOMIC VALUE AND THE REGIONS

Trade theory and the economics of location were the foundation-stones of regional science (Meyer, 1963), which in turn has provided the conceptual framework for planning over the last twenty years (Friedmann and Weaver, 1979).[6] Free Trade within the national economy presumably could be taken for granted, and the ideas of comparative advantage and geographic specialization became the main components of regional science arguments (see Chapter 5). In the sparsely settled environment of North America Harold Innis (1930) had identified *staple exports* as the basis of capital accumulation and eventual industrialization. Douglass C. North (1955) adapted this *export base* concept as the primary explanation for regional economic development, and Perloff *et al*. (1960) used it as the key to their classic empirical analysis of *Regions, Resources and Economic Growth*. Whatever the historical reality—and there can be no doubt that resource exploitation for the market has played a formative role in North American life—economic base theory became the *idée fixe* of regional science. Regional growth models from North through Borts and Stein (1962) and Richardson (1973a) have rested on arguments concerning comparative advantage, specialization, interregional factor flows, and eventual diversification. Interpretation of input/output models and such techniques as industrial complex analysis have all been directly dependent upon this singular view of economic value creation. In addition, regional planning strategies of both the *areal development* (Klaassen, 1965) and *growth centres* (e.g. Kuklinski, 1972;

Kuklinski and Petrella, 1972) schools discussed in Chapter 5 have been explicitly dedicated to encouraging specialization as the royal road to development.

Looking back to Figure 8.1, the logic of such constructs was quite simple. If the area bounded by the broken line represents a region, inside its borders will be found some combination of labour power (LP), physical resources (PR), and capital (K). Development was said to be dependent upon putting together a factor mix which could produce a specific commodity (PEV) at a competitive price which could be sold on the national or international market—realizing exchange value (REV'). All else would flow from this beginning. If additional productive forces were necessary they could be brought in from the outside. Typically this meant bringing in exogenous capital (K'), but resources (PR') or labour (LP') could also be supplied if the other inputs seemed attractive enough. Exchange value gained through export (REV') could then be used to make complementary local investments, building up the regional economic base and leading to a Rostowian world of sustained economic growth (Rostow, 1961).

As we saw in Chapter 6, the success of regional policies in France, Britain, and the United States founded on such thinking has been problematic. Even liberal commentators give them doubtful marks (e.g. Richardson, 1978a), while a burgeoning school of critical analysts point to their fundamental structural fallacies and ideological motivation (e.g. Carney, Hudson, and Lewis, 1980; Editorial Collective, 1978; Holland, 1976a; Lipietz, 1977). The two major variants of this radical critique, outlined in Chapter 7, can also be interpreted in terms of the relationships shown in Figure 8.1.

In the last chapter we saw that *underdevelopment/dependency theory* (UDT) had its origins in the disillusionment of Third World intellectuals with the outcomes of capitalist development (Frank, 1967). The adaptation of its fundamental indictment of neo-colonialism and the multinational corporation was also traced as an explanation of regional problems in industrialized countries (Hechter, 1975; Holland, 1976a). UDT begins with the observation that direct investment by the multinational corporation or meso-sector (K) gives corporate decision makers the capability of organizing other regional production inputs (i.e. LP, PR) to suit their own specialized purposes, thus controlling the region's potential exchange value (PEV) and internalizing the exchange value realized through 'domestic' sales and exports (REV, REV'). This decreases the sphere of use value production (UV) as well as the potential exchange value (PEV) which can be produced by existing capital (K). Local capital accumulation is short-circuited, and specialized production for export imposes significant opportunity costs on the regional economy.

The outcome of this process, as portrayed by some UDT writers (Hechter, 1975; Lafont, 1971), is a cultural division of labour—internal colonialism. Without the attributes of sovereignty, peripheral regions are unable to promote local accumulation (K) and the development of a well-rounded local economy. Dependent institutions lay them open to a continuing cycle of economic specialization and underdevelopment at the hands of national and multinational capital (K′). Producing only the most limited range of regionally oriented commodities (PEV), little exchange value can be realized locally (REV) and corporate decision makers can choose at any time to cut off the slim exogenous sources of regional livelihood (REV′). (Indeed, despite their attempts at negotiation and appeasement, many nation-states today find themselves in a similar position.) Group solidarity and regionalist activism in Brittany, Scotland, and Wales, for example, are perhaps inevitable sequels in this chain of events, striking at the dependent institutions which restrict more effective forms of political action.

The second strand of radical regional science reviewed in Chapter 7 takes its roots in Marxist economic theory and provides a more generalized analysis of uneven development under monopoly capitalism (e.g. Mandel, 1978). The health of a regional economy can be explained by an understanding of *capital circuits* (Palloix, 1977) and the contemporary *geographic organization of production* (Massey, 1976). For illustrative purposes Figure 8.1 shows only the simple production (PEV) and realization of exchange value (REV, REV′) through commodity sales. In fact, the self-expansion of capital occurs in all three capital circuits: commodity, finance, and production. Each time capital is transformed from one circuit to another surplus value is expanded and new wealth comes into being. Today, in the framework of the multinational corporation, self-expansion takes place on a world scale and the predominant capital circuit is the sphere of production itself (Palloix, 1977). Not only do the multinationals increase their holdings by monopoly control of globalized markets and auto-financing, but each new export platform scattered across the world periphery increases their internalized stock of productive capital (K′) and potential exchange value (PEV). While monopoly capital benefits at each step along the way, regional populations gain few rewards in terms of economic rent from export sales (REV′), interest and equity shares from finance (K), or increased productive capacity (K) and inventories (PEV).

This 'divide and conquer' strategy is implemented through the global organization of geographic specialization by *production stage* (Massey, 1976). Each region is allocated one specific operation to accomplish which is then duly passed on within the corporate structure to some other location for further assembly and processing. Few branch plants produce commodities for final demand; they typically deal in components. Frequently there is no local exchange value that could even be realized (REV) from their activities, only an accounting entry forwarded to the corporate headquarters—as are all other forms of capital expansion which take place within the corporate structure.

Running throughout the skein of underdevelopment/dependency theory and Marxist regional science is a focus on commodity relations, exchange value, and capital circuits exploded to the global scale. This is an accurate reflection of dominant forces in the contemporary economy. It provides critical analysis with poignant insights. Concomitantly, however, like liberal regional scientists and the classical economists, radicals have built a theoretical framework based almost exclusively on the analysis of production for exchange and export (REV'). We should now ask ourselves, however, whether an alternative perspective could provide a practicable starting point for regional development.

A tentative answer must begin with the fundamental observation, drawn from the arguments of economists in the classical tradition and advanced in the preceding discussion, that the sources of wealth in a territorial economy simply are not limited to exchange value realized through exports (REV'). The emerging middle class under mercantile capitalism did indeed owe its position to the capture of such wealth; industrial capitalism organized its production; and the contemporary multinational corporation has elevated its manipulation to the level of genuine wizardry. But these historical realities cannot be allowed to circumscribe our field of vision. The actuality is far more complex.

3. MORE METAPHYSICS: MARKET PRICE, MARGINAL UTILITY, AND USE VALUE

The classical distinction between use value and exchange value was a necessary analytical device to further our understanding of capitalist accumulation. Smith, Ricardo, J.S. Mill, and Marx all recognized it, Marx making it the starting point of his whole analysis. In the introductory chapter of *Capital* (1887, p. 36), Marx quoted John Locke (1691) to the effect that use value or 'The natural worth of anything consists in its fitness to supply the necessities, or serve the conveniences of human life'. Turning to exchange value, he went on (p. 39) to cite an anonymous author writing around 1740: 'The value of them (the necessities of life), when they are exchanged the one for another, is regulated by the quantity of labour necessarily required, and commonly taken in producing them.' Again, citing Benjamin Franklin (1826), 'Trade in general being nothing else but the exchange of labour for labour, the value of all things is . . . most justly measured by labour' (p. 51). Thus, use value becomes a unique quality of concrete objects, defined by biological, social, and psychological 'necessities and conveniences', while exchange value is a generalized measure of quantity: '. . . no more than exchanging one man's labour in one thing for a time certain, for another man's labour in another thing for the same time' (p. 47). Abstract labour power embodied in commodities provides the universal 'ether' which constitutes the economic value of things that enter the marketplace. When a universally accepted symbol or placeholder such as *money* is assigned to embodied labour power, its economic value may be freely traded and accumulated. It is here that we find the starting point of almost all mainstream economic theorizing since Alfred Marshall—not with a theory of 'absolute value', but with a supposedly practical theory of *money value*, based on market price (Marshall, 1890).

While we must admit, with Phyllis Deane (1978), to be exploring the realm of pure metaphysics, this paradigm shift—from the classical tradition to 'neoclassicism'—is of the most central importance to regional theory. The leap from searching for measures of absolute economic value—Ricardo's albatross—to shelving the question entirely in favour of the (micro-economic) consideration of supply and demand, marginal utility, and market price, shifted attention from the macro-economics of political economy to the mathematics of stable equilibrium. It explicitly collapsed the distinction between use value and exchange value, while implicitly redefining the source of economic wealth, swinging it from labour power to capital. Despite Marshall's claim that market price necessarily reflected production costs through its impact on the marginal utility calculus, money became value-in-itself.[7] The question of 'positive instability', viz. growth, was divorced from the process of *creation* through economic production ahd hinged on *acquisition* by earning other people's money.

From the standpoint of regional theory, whether marginal utility thinking approximates typical market choice mechanisms, whether assigned marginal utilities reflect a potential consumer's use value appraisal of commodities, and whether market price, in turn, reflects an equilibrium between production costs (including labour power) and utility are probably tangential questions. This line of reasoning has one overwhelming characteristic, however; it is *money-centred*. It equates use value with exchange value and argues that nothing more need reasonably be known about the latter than its equivalence to market price. In vulgar usage: valuation causes price and necessarily, pushed to the extreme, price causes valuation.

This may well be true for Thorstein Veblen's 'conspicuous consumption' and more generally for the Madison Avenue world view, but it is certainly a derived logic (Veblen, 1899). Descriptive of a world of distribution and circulation: salesmen, accounting offices, bankers, and financiers. Pointedly, it says nothing about the processes of economic production or the qualitative aspects of human needs. These are a matter of combining human energy and labouring skills with resources to create things which can be put to specific human uses. In the abstract they can be translated into exchange value, which is undoubtedly impacted independently by distribution mechanisms such as money and the market, but let us concentrate our attention here on the rude process of use value production.

4. THE 'RETREAT TO SUBSISTENCE': A MODEST UTOPIA

At the most basic level the regional development problem is usually a matter of incrementally improving a regional community's material standard of living; many other issues may flow from this beginning, but most typically this is the beginning point. On closer inspection, a 'low standard of living' is merely social science shorthand for an identifiable group of specific 'use-expectations' which are not being fulfilled. These are commonplace things, defined by historic cultural patterns, class relationships, and habitat. They can be enunciated and listed. Such a list would match up concrete use-needs with particular population groups for whom they are salient (e.g. social classes, age cohorts, neighbourhoods). The problem then becomes one of identifying the skills and resources available to create the use values these people feel are lacking. Things that cannot be reasonably or adequately produced by their own labour and the resources available to them have to be obtained through wider use value production processes or some form of exchange. It will be noted that this approach does not, in fact, begin with nominal concepts such as 'standard of living', 'money', and 'the market', but rather, substantively, with an assessment of concrete use needs and an inventory of the ways in which they might be met, starting with the most straightforward potential solution—do-it-yourself—and working outwards. If the need in question lies within the realm of some form of use value production or simple commodity exchange, then the problem shifts to a matter of politically organizing the necessary labour power. Conceivably this could range from organizing a household or living group to mobilizing an entire regional population.

The forms of use value and simple commodity production can be broken down into the three groupings shown in Figure 8.2. Each has its role in fulfilling different types of use-expectations. *Simple use value production*, in the first column, refers to what is often called the *household economy*. Until very recently, through most of the period dominated by urban-industrial expansion in France, Britain, and the United States, this was probably the sector of the economy which suffered the most drastic contraction (Burns, 1975).[8] First the workplace and hearth were separated one from the other under the factory system of production, as we saw in Chapter 2, and then the reproductive needs of the family were gradually subsumed within the market economy. The longstanding sexual division of labour became increasingly demeaning and senseless. 'Women's work' within the household lost both its meaningfulness and much of its utility. Even preparation of the evening meal was frequently shifted to the exchange sector: TV dinners and pizza-delivered-to-your-door. The exodus of women into the workplace over the last two generations has been not only a matter of personal liberation and economic independence; it has become a profound economic necessity, to secure the material viability of the staggering nuclear family as it has been absorbed more and more completely into the market economy.

There have been significant counter trends as well during the last quarter century which suggest a major revival in household production (Stretton, 1978).[9] A revival, without the earlier sexual reference points, which appears to be moving beyond the stage of minimal reproductive needs. It is here, I would argue, that a radically different line of regional development thinking can find its roots—building from the bottom up.

In the single-person household *production for one's self* (1A in Figure 8.2) becomes an all-important task (which must be understood in the context of living-group production discussed immediately below), but in the broader living group it is a more limited responsibility. Its important components probably include the most intimate tasks: education through study, personal hygiene, creative activity, solitary recreation, self-reflection, and the like. These activities have few immediate physical outputs but undoubtedly form the

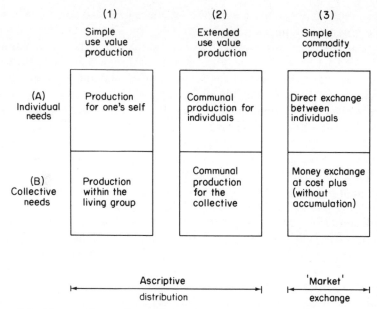

Figure 8.2   Forms of use-value and simple commodity production

basis for satisfactory participation in other group-oriented household activities. Transformation of the social fabric rests necessarily on a two-way interaction between individual strivings and group support and enablement.

Production within the living group (1B), as Le Play understood, is the key to regional economy. The ability of people living together to provide for their own daily needs sets the parameters within which the market economy will affect their lives.[10] It is at the level of the household unit that cyclical problems and structural dislocations in the regional economy are ultimately experienced, and it is here, as well, that the incentives and potential for increased self-reliance can be more effectively encouraged.

The initial problems are threefold. First, the home must be looked upon not only as a place of residence but also as a workplace. Second, the means of production (i.e. the home, land, and the necessary tools) must be effectively controlled by those who will live and work there. This probably means fee simple ownership in most situations, but in some instances even squatting and indoor or roof-top gardening have proven feasible. And third, household members must have the necessary inspiration, knowledge, and skills to fulfil many of their own reproductive and productive needs. These are no small battles in themselves, but are becoming increasingly salient for a growing number of people in all three countries. The 'urban peasant' (and more literally the rural/urban fringe peasant), with their eatable landscaping, home brew, and home-baked bread is a universal phenomenon—even a status lifestyle. Home food production, conserving, and preparation is a first small step in overcoming the contradictions between town and countryside, the exploitation of wage employment, and the rigidity of the social division of labour—a veritable greening of megalopolis, a revolutionary Victory Garden.

Energy production (mostly wood burning and solar) and the actual manufacture of things such as pottery, furniture, and clothing, as well as the reduction of transportation needs are the next steps. Home repair and building construction come in here too. People in the inner-city areas of Glasgow and Lyons, just as the exurbanites in the hinterlands of Seattle and Los Angeles, need sound places to live and work. Many are rebuilding deteriorating dwellings; some are starting afresh, splitting their own logs and making their own shakes or adobe. But this leads us quickly into a broader realm of social cooperation, which in Figure 8.2 is called *extended use value production.*

Communal production for individuals (2A) probably comes first and is the most common today. It is tied closely to the neighbourhood power movement (see Hollister, 1981; Morris and Hess, 1975; *Social Policy*, 1979) and can be divided into two categories: (1) 'barn raising' and (2) neighbourhood chores. Barn raising, taken from the historical experience of rural freeholders in Britain and France, as well as the oft-cited examples in the United States, simply means getting together to help the neighbours with tasks you may later need help in doing too: construction and harvesting for instance. Barn raising is the very essence of freely entered cooperative economy, but it is periodic in nature, seldom an ongoing task. Neighbourhood chores are continuing jobs that people do together to benefit from more muscle power—scale economies if you will. They can be productive activities such as communal gardening, which has proven highly successful as a concrete task for community organizing and spinning-off other kinds of cooperative activities (Lewis, 1979), or services like cooperative childcare—almost an imperative for low-income and many single-parent urban households.

Communal production for the collective (2B) involves the provision and maintenance of community- and region-serving infrastructure and services. This is very similar to what is done conventionally, financed through rate paying and performed by local government, with one big difference; it is done through *contributed labour power.* 'Volunteer' services have a longstanding tradition in all communities; they can and are being reinvigorated in many areas. Construction activity can also draw upon the skills and labouring ability of community residents. Their know-how has frequently contributed in times past, and once again this approach is undergoing a minor renaissance. 'Fiscal crisis' can only strike those who are dependent upon the fisc.

The most important points, perhaps, about all four components of use value production shown in columns 1 and 2 are the following. First, their creation depends on the motivation and organization of skilled human labour around explicit, agreed-upon tasks. As well, it requires, second, access to the necessary tools and resources. Finally, third, *distribution of use value* products within the concerned groups is done on an ascriptive basis, as suggested at the bottom of Figure 8.2. Everyone involved in the productive process has both rights and responsibilities which are determined through political consensus and action.

Everyone necessarily contributes to use value production and everyone benefits. There are certainly potential contradictions in such methods of productive
organization, as critics are quick to point out, *but* these are *not* the basis of
current regional problems and can only be dealt with in their turn.

It seems clear that the realm of use value production holds many possibilities
for satisfying specific human needs; it also has obvious limitations. These are
probably defined most typically by technical competence and resource availability. The fruits of specialized labour and special access to tools and
resources, however, can be incorporated in the regional economy through
simple commodity production, shown in the right-hand column of the diagram.

First of all, there is the 'grey market' for professional and other skilled
services, which has been so widely reported in all three countries in recent
years: *direct exchange between individuals* (3A). Plumbers exchange services
with dentists, bricklayers, doctors, seamstresses, and so on, in order to avoid
the inordinate fees and bottlenecks within the conventional cash economy and
the sales and income taxes imposed by government. Often, willing participants
in such schemes are now listed in local 'resource directories'—rather like
looking in the commercial section of the telephone directory, but finding here
people who might be willing to exchange a tooth extraction for having the
gutter cleaned, or maybe even a sack of green beans and some preserved fruits.
The scope of such exchanges can be significantly broadened by organization
into cooperative producer/consumer groups that, recalling Proudhon's labour
exchange, operate through a central accounting system in which labour contributions are noted and can be used as a standard unit of exchange for other
desired goods or services. The now-conventional bankcard model demonstrates the practical efficiency of such a reporting system (see the Appendix).

Here, however, we have come very close to recreating the local market
economy. The major differences are the accounting units used and their
broader implications. *Money exchange at cost plus* (3B) goes the next step,
taking us into the world of every manner of cooperative production and
distribution (see, for example Carney and Shearer, 1980; Case and Taylor,
1979; Thornley, 1981; Morris and Hess, 1975). This can include everything
from coop groceries to the linked farms and repair shops that are transformed
over time to actual manufacturing. The prime qualification for entry into this
sector is initial willingness to deal in the cash economy, *while foregoing
accumulation motives*. The massive operations of many conventional cooperatives in both Europe and North America demonstrate not only their potential
success but also the likelihood of their eventually being reabsorbed as competitors in the capitalist economy (Morris and Hess, 1975). If nothing more, they
serve to fill the gaps in the regional economy, *Perhaps leading to renewed
production for export as well as serving regional needs*.

5. SOME PERTINENT THOUGHTS ON ECONOMIC LOCATION

Moving out of the use value sphere, then, where can we reasonably expect this latter transformation to be successful? The standard economic parameters of success, at least, steer us back to the location theory debate reviewed in Chapter 7. The redeployment of capital has achieved, or, better, is achieving, a relatively stark functional reordering of the industrial map. Agglomeration economies have yielded to a labour 'cost/skill' calculus as the cardinal principle of locational decision making.[11] Globally extended oligopolists locate various production processes on the basis of minimizing the required labouring skills and, thus, labour costs at each stage of production. Technological development and global market control are the prime determinants of success. Aggregate cost minimization on a worldwide balance sheet ultimately yields corporate profit maximization (Rees, Hewings, and Stafford 1981).

An important distinction, however, must be made between the technological and capital requirements for entry into production and the market control and organizational requirements of market entry. Both sets of factors play a crucial role in global corporate operations, but they also point to the limits of multinational dominance.

The new spatial division of labour was accomplished, in part, through access to large-enough reserves of money capital to achieve long-term control of substantial national and international markets and the financial liquidity to weather short-term price wars and buy out competitors. This is not an unambiguous phenomenon, however, and we will return to it shortly. Hand in hand with the use of financial power in corporate competition large money reserves have allowed multinational firms to foot the bill for research and development, which in turn has aided them in taking maximum advantage of new factor mixes and production possibilities.

This cycle of R&D, product innovation, and radically altered production methods has had two results which are central to the regional development problem. First, with the increasing sophistication of fabricating techniques and the historical maturity of product cycles, production processes have been standardized and de-skilled (Vernon, 1966). Second, based on these changes, labour requirements—even in a highly technical age—have frequently become less demanding, lowering the wage bill and permitting a worldwide search for cost-efficient locations where 'externalities' such as unionization and government regulations and taxes do not interfere with cost minimization.

It may not be hyperbolic to conclude then that, given current manufacturing techniques, most production processes not dependent upon some special resource input, power source, or particularly rapid market access can be carried on almost anywhere, in terms of the labouring skills and standardized technologies required. Capital requirements for actual production have also fallen substantially, as Henry Ford's production-line methods have overflowed the integrated factory into the global workplace. The 'spatial margins of profitability' (Smith, 1981) have expanded to encompass most of the earth's inhabited regions: relatively small, prefabricated buildings, housing several dozen semi-skilled workers and standardized machines can be put up almost anywhere and function effectively (Friedmann, 1968).[12] While this technological regime is undoubtedly transient, like the other 'industrial revolutions' before it, arguably, it does define the contemporary technical and micro-economic parameters of many locational decisions.

The operative constraints to developing regional industries and engaging in independent export production seem to hinge on market control and the organizational requirements of market entry. There can be no question that a relatively small number of corporate organizations, with headquarters ever-more concentrated in a few metropolitan centres, control an increasing portion of production and trade (Stephens and Holly, 1980). This is accomplished in large measure through institutional organization and oligopolistic market conquest. The multinational corporation, with a range of product lines, discrete production-stage assembly operations, and far-flung production facilities, has the proven capability of cut-throat price competition and can dump on selected regional markets. As local competitors begin to feel the profit squeeze multi-nationals also have the requisite capital fluidity to buy out the losers, convert-ing their fixed capital to fit corporate needs or rapidly amortizing it and closing down the facilities. Once the field is thinned out and market control estab-lished, pricing patterns are changed to extract monopoly profits.

Only seldomly can regionally based competitors reestablish themselves. This is so for two reasons. First, the multinational can reembark on price wars whenever threatened by a new challenger. And second, for local firms to gain the technical efficiency of 'branch plants', in many instances they would have to establish entry into a broader production and distribution network, providing access to wider markets and a multinational-like division of labour. This would seem to require competitors to enter the field with three *uncommon advan-tages*: (1) substantial venture capital, (2) specialization *with diversity*, and (3) interregional linkages. How can such a package of 'comparative advantages' be put together in a peripheral, underdeveloped, or declining region?

The very success of multinational tactics may provide the motivation and set the groundwork for challenging their present hegemony. First of all, they have demonstrated beyond all doubt that given the proper institutional matrix relatively small-scale factories can operate efficiently within broadly defined spatial limits. Yet at the same time, in the process of gaining market control and rationalizing corporate production the multinationals have also initiated a downwards cycle of deindustrialization in many areas, creating a barren econ-omic landscape—a socioeconomic desert. The lack of local linkages and multi-pliers in most regions prohibits reforestation or secondary growth. This blatant contradiction, built into the very structure of multinational capitalism, provides an entry point for regional reconstruction—an entry point which can be articu-lated with the use value and simple commodity production activities described earlier to recreate the fabric of a whole, unrent regional economy.

The three 'uncommon advantages' required to underwrite such a strategy are not necessarily out of reach. Cooperative development banks and special regional development funds are already operative in many parts of the world and can provide much of the needed capital funds to construct and sustain modest-sized production facilities and infrastructure.[13] In a cooperative political approach, money capital can be supplemented by donated and socially required labour power and skills already activated within the use value economy. Specialization with diversity is also attainable given the necessary political commitment. Stephen Hymer (1972) suggested the mechanism a number of years ago: the *regional corporation* (or 'anti-multinational'), which organizes several different lines of economic activity on a territorial basis, rather than vice versa. Finally, the ability to progressively adapt an efficient spatial division of labour and gain access to international markets can be achieved through *interregional social contracts* between regional corporations and other production entities, similar to those advocated by Proudhon over a century ago (see Chapter 3). Today the objective conditions may simply exist for realization of the idea.

The tasks of regional planning in this scenario are to help identify the concrete possibilities for regional production and bring these to popular attention. Political organization and mobilization around these jobs come next, and then the creation and articulation of regional institutions to accomplish them.

NOTES

1. With due consideration for vocabulary and theoretical orientation, this position seems consistent with the classic 1960 analysis of Perloff *et al.*, *Regions, Resources and Economic Growth* (see Chap. 5, pp. 63–74). It might be argued that in practical terms the main evolutionary change in regional science perspectives over the last 20 years has come in the *political realm*. While liberal political economists like Perloff, following Tugwell's institutionalist view, believed that a right-minded national government could and should intervene in regional problems 'to achieve the important objective of rapid increases in *family* levels of living, particularly for those currently in the lower income groups' (p. 607), many radical regional scientists today consider this an unlikely turn of events. The capitalist state, immediately concerned with encouraging efficient capital accumulation, tends to safeguard existing social relations through minimal transfer payments. If the French, British, and American regional policies reviewed in Chapter 6 provide a reasonable indication, the more contradictory task of genuinely stimulating regional development—'side-tracking' investment capital and creating effective alternative political voices in the periphery—may be an unreasonable expectation from central government.

2. Considering the deteriorating UK situation, Dunford, Geddes, and Perrons (1980) frame their argument in terms of convergence through poverty sharing rather than income convergence through new growth, which as we saw in Chapter 5 and note 1 above was part of the argument of Perloff *et al.* (1960) twenty years earlier.

3. I would like to thank Michael Dear, William Goldsmith, Torsten Hägerstrang, Michael Hebbert, Ray Hudson, Jim Lewis, René Parenteau, and Ernest Weissmann for their detailed criticism of an earlier version of this argument.

4. The well-known political demands of the bourgeoisie throughout the age of revolution were based on this very programme; see, for example, Hobsbawn (1962) and Rudé (1964).

5. Rudé (1964) observed this uneven relationship between England and France at the beginning of the industrial revolution:

> Pitt, the Prime Minister, who had read Adam Smith and been convinced by many of the

arguments of the new school of political economy, set about the work of peaceful reconstruction in vigorous style: he increased the annual revenue, reduced the national debt, kept
public expenditure on a tight string, and even signed a highly advantageous 'Free Trade'
agreement (the Eden–Vergenees Treaty of 1786) with the French: thus manufactures
prospered (pp. 59–60).

This was only in England, however. Writing of the economic crisis of 1787–1789 in
France, Rudé noted:

> From agriculture it (the crisis) spread to industry; and unemployment, already developing
> from the 'Free Trade' treaty of 1786 with England, reached disastrous proportions in Paris
> and the textile centres of Lille, Lyons, Troyes, Sedan, Rouen and Rheims (pp. 73–74).

6. Walter Isard, the father of regional science, began his career, for instance, as an
   international trade theorist.
7. See Marshall's revealing critique of Ricardo's value theory in Appendix I of the
   1948 edition of *Principles*.
8. See Scott Burns (1975) for an historical account of the transformations of household production.
9. Stretton (1978) estimates that one-third of total economic production in industrialized countries like France, Britain, and the United States is now created within the
   household economy. For an extreme case, see Freed (1980). Increasingly popular
   non-commercialized approaches to do-it-yourself production are outlined in
   Farallones Institute (1979).
10. Dolly Freed points this out poignantly with a rhetorical question:

> Have you read John Steinbeck's *The Grapes of Wrath*? . . . All that starvation, squalor,
> general misery the Okies were forced to endure stemmed from only two roots: (1) the fact
> that they didn't own their homes outright, and (2) their mule-headed determination to rely
> on the money economy. They would have had problems, but not all the grief they had, if
> they had owned their homesteads in Oklahoma. The geek who stayed behind living on wild
> rabbits probably wound up living better than anyone else in the story (Freed, 1980, p. 158).

11. Other locational principles such as market or resource orientation still remain important in
    many industrial activities, however. See McCarty and Lindberg (1966) for a classic résumé.
12. It has long been observed that even in large integrated corporations individual production
    units are relatively small (see, for example, Dore, 1973).
13. Yugoslavia, for instance, allows local communes or *opcina* to retain approximately 60 per cent.
    of locally produced revenues within the local jurisdiction. Much of this is then devoted to
    regional development (see Weaver 1981a). For a sample of the coop banking debate in the
    United States, see Shearer (1979), and on development banks, see Nathan (1978).

For interested readers the following references may be helpful for developing
in greater detail some of the ideas presented here.

Self-reliance development

Clavel, P. (1982). *Opposition Planning in Appalachia and Wales*, Temple University
Press, Philadelphia.
Friedmann, J. (1981). 'Life space and economic space: contradictions in regional
development', Working Paper DP 158, School of Architecture and Urban Planning,
University of California, Los Angeles.
Friedmann, J., and Ehrhart, S. (1981). 'The household economy: beyond consumption
and reproduction', School of Architecture and Urban Planning, University of
California, Los Angeles.
McRobie, George (1981). *Small is Possible*, Harper and Row, New York.

158

Provincial Regional Forum (1981). 'The Indian community economic development corporation', West Coast Information and Research Group, Port Alberni, B.C.

Stöhr, W. (1981). 'Alternative strategies for integrated regional development of peripheral areas', Paper presented to the Second Meeting of the European Periphery Group, Lulea, Sweden, March 1981.

Stöhr, W., and Taylor, D.R.F. (Eds) (1981). *Development from Above or Below*, John Wiley, Chichester, UK.

Weaver, C. (1982). 'The limits of economism', *London Papers in Reg. Sci.*, 11, 184–202.

Political community and reconstruction

Bowles, S., Gordong, D.M. and Weiskroft, T.E. (1983). *Beyond the Waste Land: A Democratic Alternative to Economic Decline*. Anchor Press, Garden City, New York.

Boyte, H. (1980). *The Backyard Revolution: Understanding the New Citizen Movement*, Temple University Press, Philadelphia.

Carney, M., and Shearer, D. (1980). *Economic Democracy*, M.E. Sharp, White Plains, New York.

Castells, M. (1983) *The City and the Grass Roots*, Edw. Arnold, London; and Univ. of California Press, Berkeley and Los Angeles.

Choukron, Jean-Marc (1983). 'Area Wide Labor Management Collaboration for Economic Development: Reflections on the US Experience'. A paper represented to *Colloque 1983—Redeploiment Industriel et Amenagement de l'Espace: Experiences Etranges et Realities Quebecoises* Faculte d' Amenagement, Universite de Montreal, 23–24 September, 1983.

Ergan, L., and Laurent, L. (1977). *Vivre au Pays*, Editions le Cercel d'Or, Les Sables-d'Olonne, France.

Friedmann, J. (1980a). 'Urban communes, self-management, and the reconstruction of the local state', Working Papers, School of Architecture and Urban Planning, University of California; Forthcoming in *Theory and Society*, Los Angeles.

Friedmann, J. (1980b). 'On the theory of social construction: an introduction', Working Paper DP 138, School of Architecture and Urban Planning, University of California, Los Angeles.

Gran, G. (1983) *Development by People: Citizen Construction of a Just World*, Praeger, New York. See esp. the very extensive 113 page bibliography.

Illich, Ivan (1981). *Shadow Work*, Marion Boyars Publications Inc., Salem, New Hampshire.

Shapiro, B.Z. (1977). 'Mutual helping: a neglected theme in social work practice and theory', *Canadian Journal of Social Work Education*, 3, 33–44.

Smith, C.J. (1978). 'Self-help groups and social networks in the urban community', *Ekistics*, 45, 106–115.

Smith, C.J. (1980). 'Social networks as metaphors, models, and methods', *Progress in Human Geography*, 4, 500–524.

Smith, C.J. (1981). 'Urban structure and the development of natural support systems for service-dependent populations', *Prof. Geogr.*, 33/4, 457–465.

Devolution and federalism

Bogdanor, V. (1979). *Devolution*, Oxford University Press.

Burrows, B., and Denton, G. (1980). *Devolution or Federalism? Options for a United Kingdom*, Macmillan, London.

Hacker, A. (Ed.) (1964). *The Federalist Papers*, Washington Square Press, New York.

Horvat, B., Markovic, M., and Supek R. (Eds) (1975). *Self-governing Socialism: A Reader*, International Arts and Science Press, White Plains, New York.

Lafont, R. (1976) *Autonomie de la Région à l'Autogestion*, Gallimard, Paris.
Standing Conference of Towns of Yugoslavia (1974). *The Yugoslav Commune*, Stalna Konferencija Gradova, Jugoslavije (KGJ), Beograd.
Voyenne, B. (1973). *Le Fédéralisme de P.-J. Proudhon*, Press d'Europe, Paris and Nice.
Weissmann, E. (1981). 'Planned development and self-management', Paper presented to the International Conference on *Local and Regional Development in the 1980s*, UN Centre for Regional Development, Nagoya, Japan, 11–16 Nov. 1981.

Criticisms of decentralized development

Hebbert, M. (1981). 'The new decentralism—a critique of the territorial approach', Paper presented to a conference on *Planning Theory in the 1980s*, Oxford Polytechnic, April 1981.
Hilhorst, J.G.M. (1980). 'Territory versus function: a new paradigm?', Institute of Social Studies, The Hague.
Marx, K., Engels, F., and Lenin, V.I. (1972). *Anarchism and Anarchosyndicalism*, Progress Publishers, Moscow.

Some Reflections on Local 'Labour Exchange Rights'

How might a cooperative labour exchange work in practice? I suggest the notion of labour exchange rights (LERs) as the basis for an institutional mechanism which would allow 'normal local trade' without the polarization tendencies inherent in monetary exchange or the clumsiness of barter.* Labour exchange rights provide a simple operational device at the local level that gives residents exchange power without requiring money payment. It takes some portion of their labour power as well as the commodities in question out of the marketplace, as presently understood, and recognizes the need to establish equitable local trading patterns which are not impacted by national or international circuits. The viability of such a tiered exchange system is testified to by the existence of healthy national markets in countries which use 'soft' currency, of little use on the international market. But the LER concept goes a step further, controlling the extraction and accumulation of the social surplus by individuals and groups both inside and outside the regional economy.

LER is essentially an accounting system which can be regulated by a Regional Exchange and Development Bank and actually administered through chartered credit unions, linking both the workplace and households into a LER pool. There would undoubtedly be some legal and interpretive problems at the outset because of the jealous control of currency and taxes by national governments. But these objectives need not be fatal. Local tax tokens have long invaded the central state's monopoly on legal tender, while the advantages of gaining donated work for public purposes and combating currently explosive rates of unemployment might eventually open even intransigent national governments to negotiation.

Under the LER system each adult member of the community has local LER privileges. This takes the form of possessing a type of cheque book or credit card. Goods and services produced within the region can then be exchanged

---

*Normal local trade suggests a much lower degree of labour specialization than is now typically the case under corporate hegemony and a much higher degree of self-sufficiency. People would engage frequently enough in trade, but as with the immemorial pattern of household economy first they would try to fulfil their own needs themselves.

within the region by the producer and consumer double-endorsing a LER cheque or completing a credit card invoice and turning it in, in duplicate, to their respective credit unions. The transaction is recorded in LERs calculated on the basis of standard labour units (SLUs). SLUs are very much like the per job work units used to assess service fees by everyone from mechanics to physicians nowadays. But the regional catalogue of SLUs per labour task performed and per good or service entering the regional exchange market are settled upon politically by a convention of regional interest groups, the Regional Development Convention, predominated by producer and consumer unions and community council representatives. Thus, the exchange value of a person's labour and the value of a commodity in the regional exchange market are socially set, not controlled by either 'market forces' or corporate oligopoly powers. The job of setting SLU valuations under such a scheme is certainly no more cumbersome than the present price schedules and inventory controls employed by corporate wholesalers and retailers.

Each resident may then exchange his or her labour contributions to the regional social economy—rather than a salary or business profits gained through sales—for desired regionally produced goods and services. Such a system has a host of institutional implications for regional development. It can be used:

1. To encourage the use of local commodities;
2. As an exchange mechanism which prevents regional underdevelopment by financial capital;
3. As an instrument of credit to finance regional development;
4. As an encouragement to contributions to public tasks through specially awarded LERs for public service;
5. As partial remuneration to workers engaged by private industry, encouraging growth of regional productive forces (giving perhaps tax incentives to both workers and employers to participate in the system);
6. As a means of providing a minimum LER credit to all regional households;
7. As a method of controlling regional inflation in basic consumer commodities.

The list could go on almost indefinitely.

The normal problems of controlling a black market in socially controlled economies are largely avoided here, because there is already a legal dual economy. All goods, services, and salaries and wages (except donated public service remuneration) could be paid in either legal tender or LERs. Obviously much would still have to be taken in national currency, but some portion could and would be taken in LERs, because of regional tax incentives and the clear advantages of inflation control to both producer and consumer. Participating producers have no legal right of preference between the type of exchange mechanism used in an individual transaction, and, of course, LER values do not fluctuate with market variations. Producers withholding goods and services from LER transactions are fined by the Regional Development Convention *in currency amounts* fixed for the quantity of goods or services withheld.

Going back to the example transaction given above, a double-endorsed LER cheque or credit card invoice in duplicate would go to both the consumer's and the producer's credit unions. Producers receive LERs based on their work, not on 'sales' of their goods or services, but the double-endorsement, double-entry bookkeeping system acts as a safeguard to control the LER exchange value of a producer's goods and services and to prevent LERs from being converted into cash. (A producer is awarded LERs for work performed and can exchange only the quantity of LERs accumulated, limiting the possibility of collusion between individuals to beat the system). Individual incentive to provide the necessary commodities for the community is provided by the direct relationship between time worked and LERs accumulated, but supply-side control to avoid a glut of unneeded products is accomplished through the awarding of occupational licences within the LER system by the regional convention and SLU rate revisions. Once again, this is probably no more complicated or restrictive, for that matter, than current local retail licensing and land-use zoning practices. Critical decisions, however, are made by a democratic assembly rather than a licensing bureau or planning board controlled by hegemonic interests.

The credit unions of individuals involved in any transaction send the LER cheques and credit card invoices received each day to the Regional Exchange and Development Bank which acts as a clearing house, in the manner of current commercial banks. Individuals' LER accounts are then debited and cross-checked against producers' credits for inventory/exchange verification purposes. People charging on an overdrawn account work off the debt in public service. (Large purchases can be verified by phone beforehand, as with the current bankcard system.) LERs can be saved in the same fashion as currency, although interest-bearing savings accounts are expressly unavailable, to prevent the development of a regional financial circuit in LER values. The possibility of accumulation provides an incentive for individual economy and free consumer choice within the regional exchange market, yet producers have no immediate personal incentive to sway with the demands of ability to pay, distorting local production patterns and firing inflation. Regional public and community production, consumption, and governing units can also use LERs in their exchange transactions within the region, stimulating local production without arbitrary price inflation and keeping regionally produced value within the local economy. Private producers can be enticed to participate in the LER system not only because of the local tax advantages already mentioned but also because the receipt of public services and community assistance with infrastructural needs can be expressly facilitated upon participation; this follows standard regional development policy practice, but without the normal fiscal or leakage problems.

Another innovation on the individual's side is that LERs can always be gained at the socially predetermined SLU rate by performance of needed community work. Thus unemployment can be greatly reduced and individuals have a direct *positive* incentive to add to their private exchange power by contribution to the community sphere. Productive forces are no longer entirely controlled by arbitrary private production decisions. Unemployment and underemployment are also dealt with, based on the socially recognized existence of useful community work—an apparently unending source of work. Yet shortages in supplies are avoided by the fact that after some socially defined minimum a producer's consumption power is determined by how much he or she produces.

Besides these safeguards, work can always be done for legal tender; sales can always be made in a marketplace situation for national currency. Extraregional products are always available in this fashion and can be allowed to exert variable pressure upon local production without directly limiting it, based on commodity values in LERs and the comparative value in currency prices. The owners of resources can always receive remuneration for their holdings in the capitalist marketplace, although there are marked incentives to contribute some quantity of resources to regional LER production.

Surprisingly, some might argue, there is little doubt that some measure of cooperation and support can be anticipated from traditonal regional business and Chamber of Commerce-type interests. While they would not and could not profit *directly* from LER exchange transactions, their impact upon the general health of the local economy and the growth of local productive forces would be significant and an obvious advantage for the parallel money economy.

References

Abercrombie, P. (1945). *Greater London Plan, 1944*, HMSO, London.

Addams, J. (1910). *Twenty Years at Hull-House*, Macmillan, New York.

Advisory Commission on Intergovernmental Relations (ACIR) (1972). *Multi-state Regionalism*, USGPO, Washington, D.C.

Aglietta, M. (1970). *A Theory of Capitalist Regulation*, New Left Books, London.

Ahlmann, H.W., Ekstedt, I., and Jonsson, G. (1934). *Stockholms inre Differentiering*, Stockholm.

Alcaly, R.E., and Mermelstein, D. (Eds) (1976). *The Fiscal Crisis of American Cities*, Vintage Books, New York.

Alden, J., and Morgan, R. (1974). *Regional Planning: A Comprehensive View*, John Wiley, New York.

Aldridge, M. (1979). *The British New Towns: A Programme without a Policy*, Routledge and Kegan Paul, London.

Alexander, J.W. (1954). 'The basic-nonbasic concept of urban economic functions', *Econ. Geog.*, **30** (July), 246–261.

Alonso, W. (1968). 'Urban and regional imbalances in economic development', *Economic Development and Cultural Change*, **17**, 1–14.

Alonso, W. (1971). 'The economics of urban size', *Papers of the Regional Science Association*, **26**, 67–83.

Alonso, W. (1972). 'The question of city size and national policy', in *Recent Developments in Regional Science* (Ed. R. Funck), Pion, London.

Alonso, W. (1978). 'Metropolis without growth', *The Public Interest*, **53**, 68–86.

Amin, S. (1974). *Accumulation on a World Scale: A Critique of the Theory of Underdevelopment*, 2 vols, Monthly Review Press, New York.

Anderson, J. (1981). 'Enterprise zones: a short history of a bad idea', Paper presented to the CSE Regional and Urban Working Group, Liverpool Polytechnic, 23 May 1981.

Andrews, R.B. (1953). 'Mechanics of the urban economic base: historical development of the base concept', *Land Economics*, **29** (May), 161–167.

Andrieu, J. (1971). *Notes pour Servir à l'Histoire de la commune de Paris en 1871*, Payot, Paris. Originally written during the decade of the 1870s in Paris and London.

Arcangeli, F., Borzaga, C., and Goglio, S. (1980). 'Patterns of peripheral development in Italian regions, 1964–1977', Paper presented to *The Regional Problem in Europe*, nineteenth European Congress of the Regional Science Association, London, 1979. In *Papers of the Regional Science Association*, **44**, 1980.

Arcangeli, F. (1982) 'Regional and subregional planning in Italy: An evaluation of current practice and some proposals for the future' *London Papers in Regional Science*, **11**, 57–84.

Asheim, B. (1979). 'Conceptions of space and regional development: on the fallacies of current Marxist approaches to theories of regional development', in *The Regional*

Problem in Europe (Eds R. Hudson and J. Lewis), nineteenth European Congress, Regional Science Association, 28–31 August 1979, University College, London.

Bancal, J. (1971). 'Proudhon et la commune', *Autogestion et Socialisme*, **15** (March).

Bancal, J. (1973). 'L'Anarchisme et l'Autogestion de Proudhon', *L'Europe en Formation*, **163–164**, 15–38.

Banfield, E.C. (1968). *The Unheavenly City*, Little, Brown, Boston.

Barber, W.J. (1967). *A History of Economic Thought*, Penguin, Harmondsworth.

Barlow Report (1940, 1960). Report of the Royal Commission on the Distribution of the Industrial Population, Cmnd. 6153, HMSO, London.

Barnes, W. (1982). 'Cautions from Britain: the EZ answer proves ellusive', Special Supplement to *Urban Innovation Abroad*, **6/5** (May).

Beard, C.A. (1913). *An Economic Interpretation of the Constitution of the United States*, Macmillan, New York.

Beard, C.A. (1924). *Contemporary American History, 1877–1913*, Macmillan, New York.

Begg, H.M., and Lythe, C.M. (1977). 'Regional policy 1960–1971 and the performance of the Scottish economy', *Regional Studies*, **II**, 373–381.

Bellamy, E. (1884). *Looking Backwards.* Current edition: 1959, Harper, New York.

Berry, B.J.L. (1961). 'City size distributions and economic development', *Economic Development and Cultural Change*, **9**, 573–587.

Berry, B.J.L. (1967). *Spatial Organization and Levels of Welfare: Degree of Metropolitan Labour Market Participation as a Variable in Economic Development*, US Economic Development Administration, Washington, D.C.

Berry, B.J.L. (1969). *Growth Centres and Their Potentials in the Upper Great Lakes Region*, Upper Great Lakes Regional Commission, Washington, D.C.

Berry, B.J.L. (1971). 'City size and economic development: conceptual synthesis and policy problems', in *South and Southeast Asia Urban Affairs Annual* (Eds L. Jacobsen and V. Prakesh), Vol. 3, Sage Publications, Beverly Hills.

Berry, B.J.L. (1972). 'Hierarchical diffusion: the basis of development filtering and spread in a system of growth centres', in *Growth Centres in Regional Economic Development* (Ed. N.M. Hansen), Free Press, New York.

Berry, B.J.L. (1973a). *Growth centres in the American urban system.* 2 vols. Ballinger, Cambridge.

Berry, B.J.L. (1973b). *The Human Consequences of Urbanization*, St Martin's Press, New York.

Berry, B.J.L. (1980). 'Urbanization and counter urbanization in the United States', *Annals of the American Academy of Political and Soc. Sci.*, **451**, 13–20.

Bing, A.M. (1925). 'New towns for old: can we have garden cities in America?', in L. Mumford (Ed.), 'The regional plan number', *Survey Graphic*, **54** (May). Reprinted in C. Sussman (Ed.), 1976, *Planning the Fourth Migration*, MIT Press, Cambridge, Massachusetts.

Blanchard, R. (1906). *La Flandre: Etude Géographique de la Plaine Flamande en France, Belgique, et Pays-Bas*, Armand Colin, Paris.

Bleitrach, D. (1977). 'Région métropolitaine et appareils hégémoniques locaux', *Espaces et Sociétés*, 1977, 47–65.

Bleitrach, D., and Chenu, A. (1975). 'Aménagement: régulation ou aggravation des contradictions sociales? Un example: fos-surmer et l'aire métropolitaine marseillaise', *Environment and Planning, A*, **7**, 367–391.

Bluestone, B., and Harrison, B. (1980a). *Capital Mobility and Economic Dislocation*, The Progressive Alliance, Washington, D.C.

Bluestone, B., and Harrison, B. (1980b). 'Why corporations close profitable plants', *Working Paper for a New Society*, **7**, 15–23.

Bluestone, B., and Harrison, B. (1982). *The Deindustrialization of America*, Basic Books, New York.

Boardman, P. (1944). *Patrick Geddes: Maker of the Future*, University of North Carolina Press, Chapel Hill.

Boardman, P. (1978). *The Worlds of Patrick Geddes*, Routledge and Kegan Paul, London.

Bobbio, N. (1978). 'Is there a Marxist theory of the State?', *Telos, 35*, 5–16.

Boisier, S. (1980). 'Growth poles: are they dead?', UN Latin-American Institute for Economic and Social Planning, Santiago, Chile.

Bookchin, M. (1973). 'The myth of city planning', *Liberation*, **1973** (September/October), 24–42.

Bookchin, M. (1974). *The Limits of the City*, Harper Colophon Books, New York.

Booth, C. (1902). *Life and Labour in London*, 17 vols, London.

Borts, G.H., and Stein, J.L. (1962). *Economic Growth in a Free Economy*, Columbia University Press, New York.

Bouchet, P. (1962). *La Planification Française: Quinze ans d'Experience*, Editions du Seuil, Paris.

Boudeville, J.R. (1960). 'A survey of recent techniques for regional economic growth', in *Regional Economic Planning: Techniques of Analysis*(Eds W. Isard and J.H. Cumberland), Papers and Proceedings of the First Study Conference on Problems of Economic Development organized by the European Productivity Agency.

Boudeville, J.R. (1961). *Les Espaces Economiques*, PUF, Paris.

Boudeville, J.R. (1966). *Problems of Regional Economic Planning*, University Press, Edinburgh.

Boyte, H.C. (1979). 'Neighbourhood power—a term representing a new constituency entering national life', *NY Times*, 19 August.

Braudel, F. (1973). *Capitalism and Material Life, 1400–1800*, Harper Colophon Books, New York.

Bray, J. (1970). *Decision in Government*, Victory Gollancz, London.

Brenan, G. (1962). *The Spanish Labyrinth*, Cambridge University Press, Cambridge.

Brooke, M.Z. (1970). *Le Play: Engineer and Social Scientist, the Life and Work of Frédéric Le Play*, Longman, London.

Brookfield, H. (1975). *Interdependent Development*, Methuen: London.

Bruere, R.B. (1925). 'Giant power—regional-builder', in L. Mumford (Ed.), 'The regional plan number', *Survey Graphic*, **54** (May). Reprinted in C. Sussman (Ed.), 1976, *Planning the Fourth Migration*, MIT Press, Cambridge, Massachusetts.

Brunhes, I. (1910). *La Géographie Humaine*, Armand Colin, Paris.

Built Environment (1981). Theme: 'Enterprize zones', 7/1.

Burchell, R.W., and Listokin, D. (Eds) (1980). *Cities under Stress: The Fiscal Crisis of Urban America*, Centre for Urban Policy Research, Rutgers University, New Brunswick, N.J.

Burns, S. (1975). *The Household Economy: Its Shape, Origins and Future*, Beacon Press, Boston.

Business Week (1980). Special issue: 'The reindustrialization of America', No. 2643 (30 June).

Butler, S.M. (1981). *Enterprise Zones: Greenlining the Inner Cities*, Universe Books, New York.

Butler, S.M. (1982). 'Enterprise Zone Act of 1982: the administration plan', The Heritage Foundation, Issue Bulletin No. 80 (29 March).

Buttell, F.H., and Newby, H. (Eds) (1980). *The Rural Sociology of the Advanced Societies*, Allanheld, Osmun & Co., Montclair, N.J.

Cahiers d'Economie Politique (1976). Special Issues, Vols 2 and 3, 'Problemes d'economie spatiale urbaine et régionale'.

Cameron, G.C. (1970). *Regional Economic Development: The Federal Role*, Published for Resources for the Future, John Hopkins, Baltimore.

Cameron, G.C. (1974). 'Regional economic policy in the United Kingdom', in Hansen,

N.M. (ed.) *Public Policy and Regional Economic Development: The Experience of Nine Western Countries*. Ballinger, Mass.

Carney, J.R. (1980). 'Regions in crisis: accumulation, regional problems and crisis formation', in *Regions in Crisis* (Eds J.R. Carney, R. Hudson, and J.R. Lewis), Croom Helm, London.

Carney, J.R., Hudson, R., Ive, G., and Lewis, J. (1975). 'Regional underdevelopment of the North East of England', in *Proceedings of the Conference on Urban Change and Conflict*, University of York, January 1975, Centre for Environmental Studies, London.

Carney, J., Hudson, R., and Lewis, J. (Eds) (1980). *Regions in Crisis*, Croom Helm, London.

Carney, M., and Shearer, D. (1980). *Economic Democracy*, M.E. Sharp, White Plains, New York.

Carrère, P., Catin, M., and Lamandé, J. (1978). 'Evolution de la situation économique des régions françaises de 1972 à 1977', *Economie et Statistique*, **100**, 39–50.

Carroll, P.N., and Noble, D.W. (1977). *The Free and the Unfree: A New History of the United States*, Penguin Books, Harmondsworth.

Carter, I. (1974). 'The Highlands of Scotland as an underdeveloped region', in *Sociology and Development* (Eds E. de Kadt and G. Williams), Tavistock, London.

Case, J., and Taylor, R.C.R. (1979). *Co-ops, Communes and Collectives: Experiments in Social Change in the 1960's and 1970's*, Pantheon Books.

Cash, W.J. (1941). *The Mind of the South*, Alfred Knopf, New York.

Castells, M. (1972). *La Question Urbaine*, François Maspero, Paris.

Castells, M. (1973). *Luttes Urbaines et Pouvoir Politique de l'Etat Capitaliste*, Maspero, Paris.

Castello, M. (1978). *City, Class and Power*, Translation directed by E. Lebas, St Martin's Press, New York.

Chadwick, E. (1842). *Report on the Sanitary Condition of the Labouring Population of Great Britain*, London.

Charles-Brun, J. (1911). *Le Régionalisme*, 2nd ed., Bloud, Paris.

Chase, S. (1925). 'Coals to Newcastle', in L. Mumford (Ed.), 'The regional plan number', Survey Graphic, **54** (May). Reprinted in C. Sussman (Ed.), 1976, *Planning the Fourth Migration*, MIT Press, Cambridge, Massachusetts.

Chase, S. (1933). *The Promise of Power*, John Day, New York.

Chase, S. (1936). *Rich Land, Poor Land*, Whittlsey House, New York.

Cherry, G.E. (1974). *The Evolution of British Town Planning*, Leonard Hill Books, Bedfordshire.

Childe, V.G. (1936). *Man Makes Himself*, Tavistock Press, London.

Christaller, W. (1933). *Die Zentralen Orte in Suddeutschland*, Jena.

Clark, C. (1938). *National Income and Outlay*, Macmillan, London.

Clark, G.L. (1980a). 'Capitalism and regional inequality', *Annals of the Association of American Geographers*, **70:2**, 226–237.

Clark, G.L. (1980b). 'The employment relation and spatial division of labour', Program in City and Regional Planning, John F. Kennedy School of Government, Harvard University, Massachusetts.

Clark, G.L. (1981). 'Regional economic systems, spatial interdependence and the role of money', in *Industrial Location and Regional Systems* (Eds J. Rees, G.J.D. Hewings and H.A. Stafford), J.F. Bergin, New York.

Cole, J.P. (1981). *The Development Gap: A Spatial Analysis of World Poverty and Inequality*, John Wiley, New York.

Comte, A. (1852). *Le Catechisme Positivist*. Current edition: 1966, Garnier-Flammarion, Paris.

Comte, A. (1969). *Sociologie: Textes Choisi* (Edited by J. Laubier), 3rd ed., PUF, Paris.

168

Comte, A. (1972). *La Science Sociale* (Introduced and edited by A. Kremer-Mariette), Editions Gallimard, Paris.

Conkin, P. (1959). *Tomorrow a New World: The New Deal in the Suburbs*, Cornell University Press, Ithaca, N.Y.

Conrat, M., and Conrat, R. (1977). 'How American farmers, with their horse- and steam-powered machines, became specialists—in cash and debt', *Smithsonian*, **7/12**, 48–54.

Coraggio, J.L. (1972). 'Hacia una revision de la teoria de los polos de desarrollo', *Revista Latinoamericana de Estudios Urbanos y Regionales*, **2**, 25–40. English Trans., *Viertel Jahres Berichte* (Bonn), **53**, September 1973.

Coraggio, J.L. (1975). 'Polarization, development and integration', in *Regional Development and Planning: International Perspectives* (Ed. A. Kuklinski), Sijthoff, Leyden.

Crisis Reader Editorial Collective (1978). *US Capitalism in Crisis*, Union for Radical Political Economics, New York.

Crockett, J.U. (1977). *Crockett's Victory Garden*, Little, Brown & Co., Boston.

Cullingworth, J.B. (1980). *Environmental Planning, 1939–1969*, Vol. 3, *New Town Policy*, HMSO, London.

Cultiaux, D. (1975). 'L'aménagement de la région Fos-Etang de Berre', *Notes et Etudes Documentaires*, **4164–6** (February).

Cumberland, J.C. (1971). *Regional Development: Experiences and Prospects in the United States*, Mouton, The Hague.

Daley, M.C. (1940). 'An approximation to a geographic multiplier', *Economic Journal*, **50** (June–September), 248–258.

Damesick, P. (1982). 'Issues and options in UK regional policy: the need for a new assessment', *Regional Studies*, **16/15**, 390–395.

Damette, F. (1980). 'The regional framework of monopoly exploitation: new problems and trends', in *Regions in Crisis* (Eds J. Carney *et al.*), Croom Helm, London.

Damette, F. and Poncet, E. (1980). 'Global crisis and regional crises', in *Regions in Crisis* (Eds J. Carney, R. Hudson, and J. Lewis), Croom Helm, London.

Daniels, P.W. (1969). 'Office decentralisation from London—policy and practices', *Regional Studies*, **3**, 171–178.

Daniels, P.W. (1977). 'Office location in the British conurbations: trends and strategies', *Urban Studies*, **14** 261–274.

Darwent, D.F. (1969). 'Growth poles and growth centers in regional planning: a review', *Environment and Planning*, **1**, 5–31.

DATAR (1972). 'Regional investment incentives', La Délégation à l'Aménagement du Territorire et à l'Action Régionale, Paris.

Davies, T. (1974). 'Capital, State and sparse populations: the context for further research', Paper presented to the Seminar on Marginal Areas, 24–25 Sept. 1974, University of Aberdeen Working Paper No. 18, Studies in Regional Policy, School of the Environment, University College, London.

Deane, P. (1978). *The Evolution of Economic Ideas*, Cambridge University Press.

Dear, M. (1980). 'A theory of the local state', Paper presented to the Institute of British Geographers Meeting, Lancaster.

Dear, M., and Scott, A. (Eds) (1980). *Urbanization and Urban Planning in Capitalist Societies*, Methuen, London.

De Forest, R.W., and Veiller, L. (1903). *The Tenement House Problem*, Macmillan, New York.

Delors, J. (1978). 'The decline of French planning', in *Beyond Capitalist Planning* (Ed. S. Holland), St Martin's Press, New York.

Demangeon, A. (1905). *La Picardie et les Régions Voisines, Artois, Cambrésis, Beauvaises*, Armand Colin, Paris.

de Martonne, E. (1902). *La Valachie, Essai de Monographie Géographique*, Armand Colin, Paris.

Derthick, M. (1974). *Between State and Nation: Regional Organization in the United States*, The Brookings Institution, Washington, D.C.

De Souza, A.R., and Foust, J.B. (1979). *World Space-Economy*, Chas. Merrill, Columbus, Ohio.

De Souza, A.R. and Porter, P.W. (1974). *The Underdevelopment and Modernization of the Third World*. Commission of college Geography, Resource Paper No. 28, Association of American Geographers: Washington, DC.

Dickinson, R.E. (1947). *City Region and Regionalism*, Kegan Paul, Trench, Trubner & Co., London.

Dickinson, R.E. (1964). *City and Region: A Geographical Interpretation*, Routledge & Kegan Paul, London.

Dobb, M. (1946). *Studies in Development of Capitalism*, Routledge and Kegan Paul, London.

Doherty, J.C. (1980). 'The countryside comes into its own', *Planning*, **46**, 16–18.

Domhaff, G.W. (1979). *The Powers That Be: Processes of Ruling Class Domination in America*, Vintage Books, New York.

Dore, R. (1973). *British Factory, Japanese Factory: The Origins of National Diversity in Industrial Relations*. University of California Press, Berkeley.

Dulong, R. (1976a). *La Question de Bretagne*, Ecole des Hautes Etudes en Sciences Socialies, Centre d'Etude des Mouvements Sociaux, Paris.

Dulong, R. (1976b). 'La crise du rapport etat/societe locale vue au travers de la politique regionale', in *La Crise de l'Etat* (Ed. N. Poulantzas), PUF, Paris.

Dulong, R. (1978). *Les Régions, L'Etat et la Société Locale*, PUF, Paris.

Dunbar, G.S. (1978). *Elisée Reclus: Historian of Nature*, Shoe String Press, Hamden, Connecticut.

Duncan, O., *et al.* (1960). *Metropolis and Region*, Johns Hopkins Press, Baltimore.

Dunford, M. (1977). 'Regional policy and the restructuring of capital', Working Paper No. 4, Urban and Regional Studies, University of Sussex, Brighton.

Dunford, M. (1979). 'Capital accumulation and regional development in France', Working Paper No. 12, Urban and Regional Studies, University of Sussex, Brighton.

Dunford, M.F., Geddes, M., and Perrons, D. (1980). 'Working notes on the crisis and regional policy', Paper presented to the CSE Regionalism Conference, April 1980.

Dusenberry, J.S. (1950). 'Some aspects of the theory of economic development', *Explorations in Entrepreneurial History,* **3:2**.

Editorial Collective (Eds) (1978). 'Special issue on uneven regional development', *Review of Radical Political Economics,* **10/3** (Fall).

El-Shakhs, S. (1965). *Development, Primacy and the Structure of Cities*, Unpublished Ph.D. dissertation, Harvard University.

El-Shakhs, S. (1972). 'Development, primacy, and systems of cities', *Journal of Developing Areas,* **7**, 11–36.

Emmanuel, A. (1972). *Unequal Exchange: A Study of the Imperialism of Trade*, New Left Books, London.

Engels, F. (1872). *The Housing Question*, Leipzig. Current edition: 1975, Progress Publishers, Moscow.

Engels, F. (1873). *The Bakuninists at Work*, Leipzig. Current edition: 1971, Progress Publishers, Moscow.

Engels, F. (1880). *Socialism: Utopian and Scientific*, Paris. Current edition: 1975, Foreign Languages Press.

Engels, F. (1887) *The Housing Question*. London. Current edition published in 1975 by Progress Publishers, Moscow.

Engels, F. (1892). *The Condition of the Working-class in England*, London. Current edition: 1973, Progress Publishers, Moscow.

Ewen, L.A. (1978). *Corporate Power and Urban Crisis in Detroit*, Princeton University Press.

Fanon, F. (1961). *Les Damnés de la Terre*, Maspero, Paris. English trans., *The Wretched of the Earth*, 1966, Grove Press, New York.

Farallones Institute (1979). *The Integral Urban House: Self-reliant Living in the City*, Sierra Club Books, San Francisco.

Firn, J. (1975). 'External control and regional development: the case of Scotland', *Environment and Planning, A,* **7**, 393–414.

Flory, T. (1966). *Le Mouvement Régionalists Français*, PUF, Paris.

Fourier, C. (1808, 1841). *Théorie des Quatre Mouvements*, 2nd ed., Librairie Sociétaire, Paris. Original 1808 edition published in Lyons under the pseudonom Leipsic.

Fourier, C. (1822). *L'Association Domestique Agricole*, Bassange, Paris.

Fourier, C. (1829, 1848). *Le Nouveau Monde Industriel et Sociétaire*, 3rd ed., Librairie Sociétaire, Paris.

Fourier, C. (1971). *Design for Utopia: Selected Writings of Charles Fourier* (Trans. by Julia Franklin), Schocken Books, New York. Selections originally published between 1808 and 1850.

Frank, A.G. (1967). *Capitalism and Underdevelopment in Latin America*, Monthly Review Press, New York.

Frank, A.G. (1971). *Sociology of Development and Underdevelopment of Sociology*, Pluto Press, London.

Freed, D. (1980). *Possum Living: Living Easy Off the Land Without a Job and Almost No Money*, Bantam Books, New York.

Freeman, T.W. (1961). *A Hundred Years of Geography*, Aldine, Chicago.

Friedmann, J. (1955). *The Spatial Structure of Economic Development in the Tennessee Valley: A Study in Regional Planning*, Program of Education and Research in Planning, Research Paper No. 1, University of Chicago.

Friedmann, J. (1963). 'Regional planning as a field of study', *Journal of the American Institute of Planners,* **29**, 168–174. Reprinted in J. Friedmann and W. Alonso (Eds), 1964, *Regional Development and Planning: A Reader*, MIT Press, Cambridge, Massachusetts.

Friedmann, J. (Ed.) (1964). 'Special issue on regional development and planning', *Journal of the American Institute of Planners,* **30/2**.

Friedmann, J. (1966). *Regional Development Policy: A Case Study of Venezuela*, MIT Press, Cambridge, Massachusetts.

Friedmann, J. (1968) 'An Information Model of Urbanization' *Urban Affairs Quarterly* Vol. IV, No. 2 December, 1968, pp. 235–244.

Friedmann, J. (1972). 'A general theory of polarized development', in *Growth Centres in Regional Economic Development* (Ed. N. Hansen), Free Press, New York. (Original edition: 1967, revised edition: 1969).

Friedmann, J. (1975). 'Regional planning: the progress of a decade', in *Regional policy* Eds J. Friedmann and W. Alonso), MIT Press, Cambridge, Massachusetts.

Friedmann, J. (1979). 'On the contradictions between city and countryside', in *Spatial Inequalities and Regional Development*, (Eds J. Osterhaven and H. Folmer), Nijhoff, Leyden.

Friedmann, J., and Alonso, W. (Eds) (1964). *Regional Development and Planning: A Reader*, MIT Press, Cambridge.

Friedmann, J., and Hudson, B. (1973). 'Knowledge to action: an introduction to planning theory', *Journal of the American Institute of Planners,* **40**, No. 1, 1–16.

Friedmann, J., and Miller, J. (1965). 'The urban fields', *Journal of American Institute of Planners,* **31**, 312–320.

Friedmann, J., and Weaver, C. (1979). *Territory and Function: the Evolution of Regional Planning*, University of California Press, Berkeley and Los Angeles, and Edward Arnold, London.

Friedmann, J. and Wolff, G. (1982). 'Notes on the future of the world city', *International Journal of Urban and Regional Research*, **6**.

Friedmann, R., and Schweke, W. (Eds) (1981). *Expanding the Opportunity to Produce: Revitalizing the American Economy through New Enterprise Development*, Corporation for Enterprise Development, Washington, D.C.

Fuguitt, G.V., Voss, P.R., and Doherty, J.C. (1979). *Growth and Change in Rural America*, Urban Land Institute, Washington, D.C.

Gallois, L. (1908). *Régions Naturelles et Noms de Pays: Etude sur la Région Parisienne*, Armand Colin, Paris.

Galpin, C.J. (1915). 'The social anatomy of an agricultural community', Agricultural Extension Station Bulletin No. 34, University of Wisconsin, Madison.

Geddes, M. (1979). 'Uneven development and the Scottish highlands', Working Paper No. 17, Urban and Regional Studies, University of Sussex, Brighton.

Geddes, P. (1915). *Cities in Evolution: An Introduction to the Town Planning Movement and to the Study of Cities*, Williams and Norgate, London. Current edition contained in M. Stalley (Ed.), *Patrick Geddes: Spokesman for Man and the Environment*, 1972, Rutgers University Press, New Brunswick, N.J.

Geddes, P. (1925). 'Talks from the Outlook Tower', *Survey Graphic*, **1925**. (February–September). Reprinted in M. Stalley (Ed.), *Patrick Geddes: Spokesman for Man and the Environment*, 1972, Rutgers University Press, New Brunswick, N.J.

Geddes, P., and Thompson, J.A. (1889). *The Evolution of Sex*, Scott, London.

Gillingswater, D. (1975). *Regional Planning and Social Change: A Responsive Approach*, Saxon House, Westmead.

Gillingwater, D. and Hart, D.A. (1978) *The Regional Planning Process*. Saxon House: Westmead.

Glasson, J. (1974). *An Introduction to Regional Planning*, Hutchinson, London.

Goddard, J.B. (1973). *Office Linkages and Location*, Pergamon, Oxford.

Gold, S.D. (1982). 'State enterprise zones: a policy review', Legislative Finance Paper No. 17, National Conference of State Legislatures, July.

Goldsmith, W. (1981). 'Enterprise zones in the United States: lessons from international experience', Paper prepared for a panel on planning and social change at the Annual Conference of the Association of Collegiate Schools of Planning, Washington, D.C., October.

Goldsmith, W. (1982a). 'Bring the Third World home', *Working Papers for a New Society*, **1982** (March), 24–30.

Goldsmith, W. (1982b). 'Review of Sternlieb, G., and Listokin, D., eds., 1981; New tools for economic development', *Journal of Regional Science*, **22/2**, 268–271.

Goldsmith, W., and Jacobs, H.M. (1982). 'The improbability of urban policy', *Journal of the American Planning Association*, **48/1**, 53–66.

Goldstein, A.H., and Rosenberry, J.A. (Eds) (1978). *The Structural Crisis of the 1970s and Beyond: The Need for a New Planning Theory*, College of Arch. and Urban Studies, Virginia Polytechnic Institute, Blacksburg.

Golob, D. (1983). 'A bibliography on enterprise zones, with selected annotations', UBC Planning Papers, CS//5, School of Community and Regional Planning, University of British Columbia.

Goodman, R. (1979). *The Last Entrepreneurs: America's Regional Wars for Jobs and Dollars*, Simon and Schuster, New York.

Goodwin, B. (1978). *Social Science and Utopia: Nineteenth-century Models of Social Harmony*, Harvester Press, Sussex.

Goodwyn, L. (1978). *The Populist Moment: A Short History of the Agrarian Revolt in America*, Oxford University Press, New York.

Gottman, J. (1959). *Megalopolis: The Urbanized Northeastern Seaboard of the United States*, MIT Press, Cambridge, Massachusetts.

Gough, I. (1975). 'State expenditure in advanced capitalism', *New Left Review*, **92**, 53–92.

Graham, F. (1940). *The Theory of International Values*, Princeton University Press, Princeton, N.J.

Graham, O.L. (1967) *An Encore for Reform: The Old Progressives and the New Deal*. Oxford University Press: New York.

Gras, C., and Livet, G. (Eds) (1977). *Régions et Régionalisme en France: du XVIIIe Siècle à Nos Jours*, PUF, Paris.

Gravier, J.F. (1947). *Paris et le Desert Français: Decentralisation, Equipement, Population*, Le Portulan, Paris.

Gravier, J.F. (1947, 1972). *Paris et le désert français*, Flammarion, Paris.

Great Britain (1973). House of Commons Expenditure Committee (Trade and Industry Subcommittee), Session 1973–74, *Regional Development Incentives: Report*, House of Commons Paper No. 85, HMSO, London.

Green, D. (1977). *To Colonize Eden: Land and Jeffersonian Democracy*, Gordon and Cremonesi, London.

Grossman, H.J. (1979). 'How a depressed region created a foreign trade book', *Planning*, **45/1**, 17–19.

Guiral, P. (1977). 'Rapport général', in *Régions et Régionalisme en France: du XVIIIe Siècle à Nos Jours* (Eds C. Gras and G. Livet).

Gunderson, G. (1976). *A New Economic History of America*, McGraw-Hill, New York.

Habermas, J. (1973). *The Legitimation Crisis*, Beacon, Boston.

Hall, P. (1966). *The World of Cities*, McGraw-Hill, New York.

Hall, P. (1975). *Urban and Regional Planning*. Wiley, New York, Halstead Press, New York and Penguin, Harmondsworth.

Hall, P. (1977). 'Green fields and grey areas', An address to the Royal Planning Institute Annual Conference, Chester, England, 15 June.

Hall, P. (1981a). 'The enterprise zone concept: British origins, American adaptations', Working Paper No. 350, Institute of Urban and Regional Development, University of California, Berkeley.

Hall, P. (1981b). 'Enterprise zones: British origins, American adaptations', *Built Environment*, **7/1**, 5–12.

Hall, P. *et al*. (1973). *The Containment of Urban England*, 2 vols, Allen & Unwin, London.

Hall, P., and Hay, D. (1978). *Growth Centres in the European Urban System*, Department of Geography, University of Reading.

Hall, P., Castells, M., and Massey, D. (1982). 'Debate on urban enterprise zones', *International Journal of Urban and Regional Research*, **6**.

Hansen, A.H., and Perloff, H.S. (1942). *Regional Resource Development*, National Planning Association, Washington, D.C.

Hansen, N.M. (1967). 'Development pole theory in a regional context', *Kyklos*, **20**, 709–725.

Hansen, N.M. (1968). *French Regional Planning*, Indiana University Press, Bloomington.

Hansen, N.M. (1970). *Rural Poverty and the Urban Crisis: A Strategy for Regional Development*, Indiana University Press, Bloomington.

Hansen, N.M. (Ed.) (1972). *Growth Centres in Regional Economic Development*, Free Press, New York.

Hansen, N.M. (Ed.) (1974a). *Public Policy and Regional Economic Development: The Experience of Nine Western Countries*, Bellinger, Cambridge.

Hansen, N.M. (1974b). 'Regional policy in the United States', in *Public Policy and Regional Economic Development: The Experience of Nine Western Countries* (Ed. N.M. Hansen), pp. 271–303, Bellinger, Cambridge.

Hansen, N.M. (1975). 'Regional policies in the United States: experience and prospects', in *Regional Development and Planning: International Perspectives* (Ed. A. Kuklinski), pp. 139–151, Sijthoff, Leyden.

Hansen, N.M. (1976). 'Are regional development policies needed?' *The Review of Regional Studies,* **V**, 11–27.

Hansen, N.M. (1981). 'Development from above: the centre-down development paradigm', in *Development from Above or Below?: The Dialectics of Regional Planning in Developing Countries* (Eds W.B. Stohr and D.R.F. Taylor), John Wiley, Chichester.

Hansot, E. (1974). *Perfection and Progress: Two Modes of Utopian Thought,* MIT Press, Cambridge, Massachusetts.

Hardison, D. (1981). *From Ideology to Incrementalism: The Concept of Urban Enterprise Zones in Great Britain and the United States,* Princeton Urban and Regional Research Center, Princeton University.

Harloe, M. (Ed.) (1976). *Captive Cities,* John Wiley, New York.

Harnetty, P. (1972). *Imperialism and Free Trade: Lancashire and India in the Mid-nineteenth Century,* UBC Press, Vancouver.

Hartshorne, R. (1939). *The Nature of Geography,* Association of American Geographers, Lancaster, Pennsylvania.

Harvey, D. (1973). *Social Justice and the City,* Johns Hopkins University Press, Baltimore.

Harvie, C. (1977). *Scotland and Nationalism: Scottish Society and Politics, 1707–1977,* George Allen & Unwin, London.

Hayden, D. (1976). *Seven American Utopias: The Architecture of Communitarian Socialism, 1790–1975,* MIT Press, Cambridge, Massachusetts.

Hechter, M. (1975). *Internal Colonialism: The Celtic Fringe in British National Development, 1536–1966,* Routledge & Kegan Paul, London, and University of California Press, Berkeley and Los Angeles.

Heil, B. (1978). 'Sunbelt migration', in *US Capitalism in Crisis,* pp. 87–102, URPE, New York.

Heilbroner, R.L. (1953). *The Worldly Philosophers,* Simon and Schuster, New York.

Heller, H.R. (1968). *International Trade: Theory and Empirical Evidence,* Prentice-Hall, Englewood Cliffs, N.J.

Hermansen, T. (1971). *Spatial Organization and Economic Development,* Institute of Development Studies, University of Mysore.

Heskin, A. (1980). 'Crisis and response: an historical perspective on advocacy planning', *Journal of the American Planning Association,* **46/1**, 50–63.

Hession, C.H., and Sardy, H. (1969). *Ascent to Affluence: A History of American Economic Development,* Allyn and Bacon, Boston.

Hilberseimer, L. (1949). *The New Regional Pattern: Industries and Gardens, Workshops and Farms,* Paul Theobald, Chicago.

Hildebrand, G., and Mace, A. Jr. (1950). 'The employment multiplier in an expanding industrial market: Los Angeles County, 1940–1947', *Rev. of Econ. and Statistics,* **32** (August), 341–349.

Hirschman, A.O. (1958). *The Strategy of Economic Development,* Yale University Press, New Haven, Connecticut.

Hobsbawm, E.J. (1962). *The Age of Revolution, 1789–1848,* Mentor Books, New York.

Hobsbawm, E.J. (1977). 'Some Reflections on the Breakup of Britain', *New Left Review,* **105**, 3–23.

Holland, S. (1971). 'Regional underdevelopment in a developed economy: the Italian case?', *Regional Studies,* **5**, 71–90.

Holland, S. (1976a). *Capital versus the Regions,* Macmillan, London.

Holland, S. (1976b). *The Regional Problem,* Macmillan, London.

Holland, S. (Ed.) (1978). *Beyond Capitalist Planning,* St Martin's Press, New York.

Holland, S. (1979). 'Capital, labour and the regions: aspects of economic, social and political inequality in regional theory and policy', in *Spatial Inequalities and Regional Development* (Eds H. Folmer and J. Oosterhaven), Nijhoff, Leyden.

Hollister, R. (1981). 'Neighbourhood price, neighbourhood power', *Planning*, **47/3**, 28–30.

Holloway, J., and Picciotto (Eds) (1978). *State and Capital*, Arnold, London.

Hoover, E.M. (1937). *Location Theory and the Shoe and Leather Industries*, Harvard University Press, Cambridge, Massachusetts.

Hoover, E.M. (1948). *The Location of Economic Activity*, McGraw-Hill, New York.

Howard, E. (1898, 1902). *Garden cities of tomorrow*. (originally published under the title *Tomorrow: a peaceful path to real reform*) First published in 1946 by Faber and Faber Limited. First published in this edition 1965. Reprinted 1970 and 1974.

Howard, E. (1946). *Garden Cities of Tomorrow*, Faber, London. Original published in 1898.

Howe, G. (1981). 'Liberating free enterprise: a new experiment', in *New Tools for Economic Development* (Eds G. Sternleib and D. Listokin), pp. 13–24, Center for Urban Policy Research, Rutgers University, Piscataway, N.J.

Hoyt, H. (1954). 'Homer Hoyt on the concept of the economic base', *Land Economics*, **30** (May), 182–186.

Hudson, R. (1980). 'Capital accumulation and regional problems: a study of North East England', Department of Geography, University of Durham.

Hughes, H.S. (1958). *Consciousness and Society: The Reorientation of European Social Thought 1890–1930*, esp. Chaps 2 and 10, Alfred A. Knopf, New York.

Hughes, M. (Ed.) (1971). *The Letters of Lewis Mumford and Frederic J. Osborn: A Transatlantic Dialogue, 1938–70*, Adams and Dart, Bath.

Humberger, E. (1981). 'The enterprise zone fallacy', *Journal of Community Action*, **1**, 20–28.

Hume, E., and Shannon, D. (1979). 'Statehood issues divide Puerto Rico', *L.A. Times*, 2 September.

Hymer, S. (1968). 'La grande corporation multinationale', *Revue Economique*, **19/6**, 949–973.

Hymer, S. (1972a). 'The multinational corporation and the law of uneven development', in *Economics and World Order: From the 1970s to the 1990s* (Ed. J.N. Bhagwati), Macmillan, New York.

Hymer, S. (1972b). 'The internationalization of capital', *Journal of Economic Issues*, **6**, 91–111.

Hymer, S., and Resnick, S. (1971). 'International trade and uneven development', in *Trade, Balance of Payments and Growth: Papers in International Economics in Honor of Charles P. Kindleberger* (Eds J.N. Bhagwati, R.W. Jones, R.A. Mundell, and J. Vanek), North-Holland, Amsterdam.

Hymer, S., and Rowthorn, R. (1970). 'Multinational corporations and international oligopoly: the non-American challenge', in *The International Corporation* (Ed. C.P. Kindleberger), MIT Press, Cambridge.

IMF (1978). *Government Finance Statistics Yearbook 1978*, Vol. 2, International Monetary Fund, Washington, D.C.

Innis, H.A. (1930). *The Fur Trade in Canada: An Introduction to Canadian Economic History*, University of Toronto Press, Toronto.

Isard, W. (1956). *Location and Space Economy*, MIT Press, Cambridge, Massachusetts.

Isard, W., and Peck, M.J. (1954). 'Location theory and international and interregional trade theory', *Quart. J. of Econ.*, **68**, 97–114.

James, P.E. (1972). *All Possible Worlds: A History of Geographical Ideas*, Odyssey Press, Indianapolis.

James, P.E., and Jones, C.F. (Eds) (1954). *American Geography: Inventory and Prospect*, Published for the Association of American Geographers, Syracuse University Press, Syracuse, N.Y.

Jensen, J. (Ed.) (1951). *Regionalism in America*, University of Wisconsin Press, Madison.

Jensen, H.T. (1980). 'The role of the State in regional development, planning and management', Dunelm Translations No. 5, Department of Geography, Durham University.

Jessop, B. (1977). 'Recent Theories of the capitalist state', *Cambridge Journal of Economics*, **1**, 353–373.

Johnson, D.A. (1983). 'Lewis Mumford: critic, colleague and philosopher', *Planning*, **49/4**, 9–14.

Jumper, S.R., Bell, T.L., and Ralston, B.A. (1980). *Economic Growth and Disparities: A World View*, Prentice-Hall, Englewood Cliffs, N.J.

Kashdan, S. (1981a). 'EDA projects in jeopardy', *Planning*, **47/5**, 8–9.

Kashdan, S. (1981b). 'Can states handle block grant load?', *Planning*, **47/8**, 8–9.

Kellogg, P.U. (Ed.) (1914). *The Pittsburgh Survey*, Published for the Russell Sage Foundation, Survey Associates, New York.

Kitchen, P. (1975). *A Most Unsettling Person*, Saturday Review Press, New York.

Klaassen, L.H. (1965). *Area Economic and Social Redevelopment: Guidelines for Programmes*, OECD, Paris.

Klaassen, L.H. (1967). *Methods of Selecting Industries for Depressed Areas: An Introduction to Feasibility Studies*, OECD, Paris.

Klaassen, L.H. (1968). *Social Amenities in Area Economic Growth: An Analysis of Methods for Defining Needs*, OECD, Paris.

Klaassen, L.H., and Drewe, P. (1973). *Migration Policy in Europe: A Comparative Study*, Lexington/Heath, Lexington, Mass.

Klaassen, L.H., Torman, D.H. van Dongen, and Koyck, L.J. (1949). *Hoofdlijnen van de Sociaal-economische Ontwikkeling der Gemeente Amersfoort van 1900–1970*, Leiden.

Kropotkin, P. (1899). *Fields, Factories and Workshops*, Hutchinson, London, Quotations taken from C. Ward (Ed.), *Fields, Factories and Workshops Tomorrow*, 1974, George Allen and Unwin, London.

Kropotkin, P. (1902). *Mutual Aid: A Factor of Evolution*, London. Current edition: n.d., Extending Horizons Books, Boston.

Kuklinski, A. (Ed.) (1972). *Growth Poles and Growth Centres in Regional Planning*, Mouton, Paris.

Kuklinski, A. (Ed.) (1975). *Regional Development and Planning: International Perspectives*, Sijthoff, Leyden.

Kuklinski, A., and Petrella, R. (Eds) (1972). *Growth Poles and Regional Policies*, Mouton, Paris.

Kuznets, S. (1941). *National Income and Its Composition, 1919–1938*, National Bureau of Economic Research, New York.

Laclan, E. (1975). 'The specificity of the political', *Economy and Society*, **4**, 87–110.

Lafont, R. (1967) *La Révolution Régionaliste*. Éditions Gallimard, Paris.

Lafont, R. (1968) *Sur La France*. Éditions Gallimard, Paris.

Lafont, R. (1971). *Décoloniser en France*, Gallimard, Paris.

Lagarde, P. (1977). *La Régionalisation*, Seghers, Paris.

Langlois, J. (1976). *Défense et Actualité de Proudhon*, Payot, Paris.

Lasuén, J.R. (1969). 'On growth poles', *Urban Studies*, **6**, 137–161.

Lasuén, J.R. (1974). 'National and urban development', in *Symposium on Urban Development*, Rio de Janeiro, 20–24 August 1973, Banco Nacional da Habitacao, Rio de Janeiro.

Law, C.M. (1980). *British Regional Development since World War I*, Davis & Charles, Newton Abbot, Devon.

Lebas, E. (1979). 'The evolution of the state monopoly thesis in French urban research—or the crisis of the urban question', Centre for Environmental Studies, London.

Lefebvre, H. (1968). *Le Droit à la Ville*, Editions Anthropos, Paris.

Lefebvre, H. (1970). *La Révolution Urbaine*, Gallimard, Paris.

Lefebvre, H. (1972). *Le Pensée Marxiste et la Ville*, Casterman, Paris.

Le Monde (1981). 'Les élections législatives de juin 1981', Supplément aux Dossiers et Documents du Monde, Juin 1981.

Lenin, V.I. (1917). *Imperialism, the Highest State of Capitalism*. Current edition: 1970, Foreign Languages Press, Peking.

Le Play, F. (1877–1879). *Les Ouvriers Européens*, 6 vols, 2nd ed., Tours, Paris.

Le Play, F. (n.d.). *L'Esquisse d'une Division Provinciale de la France*.

Levainville, J. (1909). *Le Morvan*, Armand Colin, Paris.

Leven, C. (1954). 'An appropriate unit for measuring the urban economic base', *Land Economics*, **30** (November), 369–371.

Levin, P. (1976). *Government and the Planning Process*, Allen & Unwin, London.

Levy, J. (1979). 'Pour une problematique: région et formation economique et sociale', *Espaces Temps*, **10–11**, 80–107.

Lewis, C.A. (1979). 'Plants and people in the inner city', *Planning*, **45/3** (March), 10–14.

Lewis, W.A. (1955). *Theory of Economic Growth*, George Allen & Unwin, London.

Lilienthal, D. (1944). *TVA: Democracy on the March*, Harpers, New York.

Lipietz, A. (1977). *Le Capital et Son Espace*, Maspero, Paris.

Lipietz, A. (1979a). 'Polarisation interrégionale et tertiarisation de la société', Paper presented at the nineteenth European Congress of the Regional Science Association, University College, 28–31 August 1979, London.

Lipietz, A. (1979b). 'Inter-regional polarisation and the tertiarisation of society', Translation by R. Firth and J. Lewis, Paper presented to *The Regional Problem in Europe*, nineteenth European Congress of the Regional Science Association, London, 1979. (In *Papers of the Regional Science Association*, **44**, 1980.)

Lipton, M. (1977). *Why Poor People Stay Poor: Urban Bias in World Development*, Harvard University Press, Cambridge, Massachusetts.

Lojkine, J. (1977). 'L'etat et l'urbain: contribution à une analyse matérialiste des politiques urbaines dans les pays capitalistes développés', *International Journal of Urban and Regional Research*, **1**, 256–271.

Lonsdale, R.E., and Seyler, H.L. (Eds) (1979). *Nonmetropolitan Industrialization*, Winston and Sons, Washington, D.C.

Lösch, A. (1954). *The Economics of Location* (Trans. of *Die Raumliche Ordnung der Wirtshaft*, 2nd ed., 1944), Yale University Press, New Haven.

Lower, A.R.M. (1933). 'The trade in square timber', *Contributions to Canadian Economics*, Vol. 6, pp. 40–61, University of Toronto Press, Toronto.

Lubove, R. (1969). *Twentieth Century Pittsburgh*, John Wiley, New York.

Maarek, G. (1979). *An Introduction to Karl Marx's Das Kapital: A Study in Formalisation* (Trans. by M. Evans), Oxford University Press, New York.

McCallum, J.D. (1979). 'The development of British regional policy', in *Regional Policy: Past Experiences and New Directions* (Eds D. Maclennan and J.B. Parr), Glasgow Social and Economic Research Studies No. 6, Martin Robertson, Oxford.

McCarthy, K.F., and Morrison, P.A. (1979). *The Changing Demographic and Economic Structure of Nonmetropolitan Areas in the United States*, Rand Corporation, Santa Monica, California.

McCarty, H., and Lindberg, J. (1966). *A Preface to Economic Geography*, Prentice Hall, Englewood Cliffs, N.J.

McCrone, G. (1969). *Regional Policy in Britain*, Allen & Unwin, London.

McDonald, I., and Howick, C. (1981). 'Monitoring the enterprise zones', *Built Environment*, **7/1**, 31–37.

MacFadyen, D. (1970). *Sir Ebenezer Howard and the Town Planning Movement*, Manchester University Press, Manchester.

McGee, T.G. (1971). *The Urbanization Process in the Third World: Explorations in Search of a Theory*, Bell, London.

Machlup, F. (1943). *International Trade and the National Income Multiplier*, Blakiston, Philadelphia.

MacKaye, B. (1928). *The New Exploration: A Philosophy of Regional Planning*, Harcourt, Brace, New York. Current edition: 1962, University of Illinois Press, Urbana.

MacKaye, B. (1931). 'Townless highways for the motorist', *Harpers Magazine*.

Mackaye, B., and Mumford, L. (1928). 'Regional planning', in *The Encyclopedia Britannica*, 14th ed., Vol. 19, pp. 71–72.

Mackintosh, W.A. (1923). 'Economic factors in Canadian history', *Canadian Historical Review*, **4**, (March), 12–25.

MacLennan, D., and Parr, J.D. (Eds) (1979). *Regional Policy: Past Experience and New Directions*, Glasgow Social and Economic Research Studies No. 6, Martin Robertson, Oxford.

Mairet, P. (1957). *Pioneer of Sociology: The Life and Letters of Patrick Geddes*, Lund Humphries, London.

Maldonado-Denis, M. (1972). *Puerto Rico: A Socio-historic Interpretation*, Vintage Books, New York.

Malizia, E. (1978). 'Organizing to overcome uneven development: the case of the US South', *Rev. of Rad. Pol. Econ.*, **10/3**, 87–94.

Mandel, E. (1975). *Late Capitalism*, New Left Books, London.

Mandel, E. (1968). *Marxist Economic Theory*, Vols 1 and 2, Monthly Review Press, New York.

Mandel, E. (1976). 'Capitalism and regional disparities', *Southwest Economy and Society*, **1**.

Mandel, E. (1978). *Late Capitalism*, Verso, London.

Mandle, J. (1978). 'The economic underdevelopment of the post-bellum South', *Marxist Perspectives*, **1:4**, 68–79.

Manners, G., Keeble, D., Rodgers, B., and Warren, K. (1980). *Regional Development in Britain*, 2nd ed., John Wiley, Chichester, Sussex.

Markusen, A.R. (1978a). 'Class, rent, and sectoral conflict: uneven development in Western US boomtowns', *Rev. of Rad. Pol. Econ.*, **10/3**, 117–129.

Markusen, A.R. (1978b). 'Regionalism and the capitalist state: the case of the United States', *Kapitalistate*, **7**, 39–62.

Markusen, A.R., and Fastrup, J. (1978). 'The regional war for federal aid', *The Public Interest*, **53**, 87–99.

Markusen, A.R., and Schoenberger, E. (1979). 'The political economy of regional development in the Western United States', Paper presented to the Regional Science Association meeting, November 1979, Los Angeles.

Marquand, J. (1979). *The Service Sector and Regional Policy in the United Kingdom*, Research Note No. 29, Centre for Environmental Studies.

Marsh, G.P. (1864). *Man and Nature, or, Physical Geography as Modified by Human Action*. Charles Scribner, New York. Current edition: 1964, Harvard University Press, Cambridge, Massachusetts.

Marshall, A. (1890). *Principles of Economics*, Macmillan, London.

Martin, C.H., and Leone, R.A. (1977). *Local Economic Development: The Federal Connection*, Lexington Books, Lexington, Mass.

Marx, K. (1847). *The Poverty of Philosophy*, Paris. Current edition: 1963, International Publishers, New York.

Marx, K. (1857a). *Pre-capitalist Economic Formations*. Current edition: 1965, International Publishers, New York.

Marx, K. (1857b). *Preface and Introduction to a Contribution to the Critique of Political Economy*. Current edition: 1976, Foreign Languages Press, Peking.

Marx, K. (1867). *Capital: A Critique of Political Economy*. Current edition: 1967, International Publishers, New York.

Marx, K. (1875). *Critique of the Gotha Programme*. Current edition: 1972, Foreign Languages Press, Peking.

Marx, K. (1909). *Capital* (Edited by F. Engels), 3 vols, Chas. M. Kerr, Chicago.

Marx, K. (1965) *Pre-capitalist Economic Formations*. International Pub, New York. (Originally published 1857.)

Marx, K. (1967). *Capital*, Vol. 1 (Edited by F. Engels), International Publishers, New York. (Originally published 1887.)

Marx, K. (1976). *Preface and Introduction to a Contribution to the Critique of Political Economy*, Foreign Language Press, Peking. (Originally published 1857.)

Marx, K., and Engels, F. (1846). *The German Ideology*. Current edition: 1970, International Publishers, New York.

Marx, K., and Engels, F. (1847). *The Communist Manifesto*. Current edition: 1970, Pathfinder Press, New York.

Marx, K. and Engels, F. (1970). *The German Ideology*, International Publications, New York. (Originally published 1846.)

Marx, K., and Engels, F. (1978). *The German Ideology* (Ed. by C.J. Arthur), International Publishers, New York.

Marx, K., Engels, F., and Lenin, V.I. (1972). *Anarchism and Anarchosyndicalism*, Progress Publishers, Moscow.

Massey, D.B. (1974). *Towards a Critique of Industrial Location Theory*, Centre for Environmental Studies, London.

Massey, D. (1976). 'Restructuring and regionalism: some spatial effects of the crisis', Paper presented to the American Regional Science Association. CES Working Note No. 479.

Massey, D. (1978a). 'Survey: regionalism—some current issues', *Capital and Class*, **6**, 102–155.

Massey, D. (1978b). 'In what sense a regional problem'?, Paper presented to the Regional Studies Association Conference, Glasgow, 31 January 1978, CES Working Note No. 479.

Massey, D.B. (1978c). 'Capital and locational change: the UK electrical engineering and electronics industries', *Review of Radical Political Economics*, **10**, 39–54.

Massey, D.B., and Meegan, R.A. (1978). 'Industrial restructuring versus the cities', *Urban Studies*, **15**.

Massey, D.B., and Meegan, R.A. (1979). *The Industrial Location Project, Final Report*, Centre for Environmental Studies, London.

Mather, B., and Mather, M. (1978). *Gardening for Independence*, Durrell Publications, Kennebunkport, Maine.

Mayeur, J.-M. (1977). 'Démocratie chrétienne et régionalisme', in *Régions et Régionalisme en France, du XVIIIe Siécle à Nos Jours* (Eds C. Gras and G. Livet), PUF, Paris.

Meier, G.M. (1953). 'Economic development and the transfer mechanism', *Canadian J. of Econ. and Political Science*, **19** (February), 1–19.

Mera, K. (1973). 'On urban agglomeration and economic efficiency', *Economic Development and Cultural Change*, **21**, 309–324.

Merrington, J. (1975). 'Town and country in the transition to capitalism', *New Left Review*, **93**, 71–92.

Meyer, J.R. (1963). 'Regional economics: a survey', *American Economic Review*, **53**, 19–54.

Michalet, C.A. (1975a). *Les Firmes Multinationales et la Nouvelle Division Internationale du Travail*, Document de Travail, Recherches pour le Programme Mondial de l'Emploi, Bureau International du Travail, Geneva.

Michalet, C.A. (1975b). *Transfert Technologique par les FMN et Capacité d'Absorption des Pays en Voie de Développement*, OCDE, Paris.

Michalet, C.A. (1976). 'Les firmes multinationales et la nouvelle division internationale du travail', Working Paper WEP 2–28/WP5, World Employment Paper Research, ILO, Geneva.

Michon-Savarit, C. (1975). 'La place des régions francaises dans la division internationale du travail: deux scenarios contrastés', *Environment and Planning, A,* **4,**
449–454.

Miller, S.M. (1978). 'The recapitalization of capital', *International Journal of Urban and
Regional Research,* **2,** 202–212. Revised edition: August 1978.

Mingione, E. (1977). 'Theoretical elements for a Marxist analysis of urban development', in *Captive Cities* (Ed. M. Harloe) pp. 89–103, John Wiley, London.

Moore, R. (1979) *Urban Development in the Periphery of Industrialized Societies.* Paper
presented to the Urban Change and Conflict Conference, Nottingham University, 5–8
January, 1979.

Morgan, K. (1979). 'State regional interventions and industrial reconstruction in post-
war Britain: the case of Wales', Working Paper No. 16, Urban and Regional Studies,
University of Sussex, Brighton.

Morgan, K. (1980). 'The reformulation of the regional question, regional policy and the
British State', Working Paper No. 18, Urban and Regional Studies, University of
Sussex, Brighton.

Moriarty, B. (1980). *Industrial Location and Community Development,* University of
North Carolina Press, Chapel Hill.

Morison, J.E., Commager, H.S., and Leuchtenburg, W.E. (1977). *A Concise History of
the American Republic,* Oxford University Press, New York.

Morrill, R.L. (1979). 'Stages in patterns of population concentration and dispersion',
Professional Geographer, **31,** 55–65.

Morris, E., and Hess, K. (1975). *Neighbourhood Power: The New Localism,* Beacon
Press, Boston.

Mosely, M.J. (1974). *Growth Centres in Spatial Planning,* Pergamon Press, Oxford.

Mounts, R. (1982). 'The "Urban Enterprise zone" hustle', *Commonweal,* **1982** (13
March).

Moynihan, D.P. (1978). 'The politics and economics of regional growth', *The Public
Interest,* **51,** 3–21.

Mumford, L. (Ed.) (1925a). 'The regional plan number', *Survey Graphic,* **54** (May).

Mumford, L. (1925b) 'Regions to live in', in L. Mumford (Ed.), 'The regional plan
number', *Survey Graphic,* **54** (May). Reprinted in C. Sussman (Ed.), 1976, *Planning
the Fourth Migration,* MIT Press, Cambridge, Massachusetts.

Mumford, L. (1938). *The Culture of Cities,* Harcourt, Brace and World, New York.

Mumford, L. (1961). *The City in History,* Harcourt, Brace and World, New York.

Mumford, L. (1966). 'The disciple's rebellion: a memoir of Patrick Geddes', *Encounter,*
27/3, 11–21.

Mumford, L. (1982). *Sketches from Life: The Autobiography of Lewis Mumford—The
Early Years,* Dial Press, New York.

Mumford, L., and Osborn, F.J. (1971). *The Letters of Lewis Mumford and Frederic J.
Osborn* (Edited by M. Hughes), Adams & Dart, Bath.

Mumford, L., and Osborn, J. (1972). *The Letters of Lewis Mumford and Frederic J.
Osborn* (Edited by M. Hughes), Praeger, New York.

Mumy, G.E. (1978–1979). 'Town and country in Adam Smith's *The Wealth of Nations',
Science and Society,* **42,** 458–477.

Myhra, D. (1974). 'Rexford Guy Tugwell: initiator of America's greenbelt new towns',
Journal of the American Institute of Planners, **40,** 176–188.

Myrdal, G. (1957). *Economic Theory and Underdeveloped Regions,* Duckworth,
London.

Myrdal, G. (1969). *The Political Element in the Development of Economic Theory*
(Trans. by P Streeten), Simon and Schuster, New York.

Nabudere, D. (1977). *The Political Economy of Imperialism,* Zed Press, London.

Nairn, T. (1977). *The Break-up of Britain: Crisis and Neo-nationalism,* New Left Books,
London.

Nathan, R.P. (1978). *Lessons from European Experience for a U.S. National Development Bank*, Council for International Urban Liaison, Washington.

National Urban League (1982). 'Can enterprise zones work for U.S.?', Office of Washington Operations, Washington, D.C., August.

Neutze, G.M. (1965). *Economic Policy and the Size of Cities*, Australia National University Press, Canberra.

Newman, M. (1972) *The Political Economy of Appalachia: A Case Study in Regional Integration*, Lexington Books, Lexington, Massachusetts.

New Towns Committee (1946). *Final Report* (The Keith Committee Report), Cmnd. 6876, HMSO, London.

North, D.C. (1955). 'Location theory and regional economic growth', *Journal of Political Economy*, **63**, 243–258.

North, D.C. (1961). *The Economic Growth of the United States, 1750–1860*, Prentice-Hall, Englewood Cliffs, N.J.

North D.C. (1966). *Growth and Welfare in the American Past: A New Economic History*, Prentice-Hall, Englewood Cliffs, N.J.

O'Connor, J. (1973). *The Fiscal Crisis of the State*, St Martin's Press, New York.

Odum, H.W. (1934). 'The case for regional–national social planning', *Social Forces*, **13**, 6–23.

Odum, H.W. (1936). *Southern Regions of the United States*, University of North Carolina Press, Chapel Hill, N.C.

Odum, H.W. (1939). *American Social Problems: An Introduction to the Study of the People and Their Dilemmas*, Henry Holt, New York.

Odum, H.W., and Jocher, K. (Eds) (1945). *In Search of Regional Balance in America*, University of North Carolina, Chapel Hill, N.C.

Odum, H.W., and Moore, H.E. (1938). *American Regionalism, a Cultural–Historical Approach to National Integration*, Henry Holt, New York.

OECD (1970). *Regional Policy in 15 Industrialised OECD Countries*, Organization for Economic Cooperation and Development, Paris.

OECD (1974). *Re-appraisal of Regional Policies in OECD Countries*, Organization for Economic Cooperation and Development, Paris.

OECD (1976). *Regional Problems and Policies in OECD Countries*, Vol. 1, *France, Italy, Ireland, Denmark, Sweden, Japan*, Organization for Economic Cooperation and Development, Paris.

Offe, C. (1975). 'The theory of the capitalist state and the problem of policy formation', in *Stress and Contradiction in Modern Capitalism* (Eds L.N. Lindberg, R. Alford, C. Crouch, and C. Offe), D.C. Heath, Lexington, Massachusetts.

Offe, C., and Ronge, V. (1975). 'Theses on the theory of the State', *New German Critique*, **6**, 137–147.

Osborn, J.F. (1918). *New Towns after the War*, National Garden Cities Committee, London. Reprinted in 1942 by Dent, London.

Osborn, F.J., and Whittick, A. (1963). *The New Towns: The Answer to Megalopolis*. Revised edition: 1969, Leonard Hill Books, Bedfordshire.

Osborn, J.F., and Whittick, A. (1969). *The New Towns: The Answer to Megalopolis*, Leonard Hill, London.

Owen, R. (1813). *A New View of Society, or Essays on the Principle of the Formation of the Human Character*, Cadell and Davies, London.

Owen, R. (1972). *A New View of Society and Other Writings*, Denton, London. Original essays published between 1812 and 1820.

Paelinck, J. (1965). 'La théorie de développement régional polarisé', *Cahiers de l'Institut de Science Economique Appliquée*, Series L, **15**, 5–48.

Palloix, C. (1975a). *L'Internationalisation du Capital: Eléments Critiques*, Maspero, Paris.

Palloix, C. (1975b). *La Crise Organique du Capitalisme: Essai sur la Crise de l'Organisation Capitaliste de la Production et du Process de Travail*, IREP, Grenoble.

Palloix, C. (1975c). *L'Economie Mondiale Capitaliste et les Firmes Multinationales*, Maspero, Paris.

Palloix, C. (1977). 'The self-expansion of capital on a world scale', *Review of Radical Political Economics*, **9**, 1–29.

Panikkar, K.M. (1959). *Asia and Western Dominance*, Allen & Unwin, London.

Parti Socialiste Français (1980). *Projet Socialiste*, Club Socialiste du Livre, Paris.

Passeron, H. (1979). 'Inflexion de la croissance et effects spatiaux: méthodes d'analyse mises en place lors de la preparation du VIIe plan', *Cahiers d'Economie Politique*, **7**. Numéro special: 'Planification urbaine et régionale, vers de nouvelles approches' (February), 165–170.

Pedersen, P.O. (1974). *Urban-Regional Development in South America: A Process of Diffusion and Integration*, Mouton, Paris.

Perloff, H.S., Dunn, E.S. Jr., Lampard, E.E., and Muth, R.F. (1960). *Regions, Resources and Economic Growth*, Published for Resources for the Future, Johns Hopkins University Press, Baltimore.

Perrin, J.C. (1975). *Le Dévelopment Régional*, Presses Universitaires de France, Paris.

Perroux, F. (1950). 'Economic space: theory and applications', *Quarterly Journal of Economics*, **64**, 90–97.

Perroux, F. (1955). 'Note sur la notion de "pôles de croissance" ', *Economie Appliquee*, **1** and **2**, 307–320.

Perroux, F. (1969). *Indépendance de la Nation*, Union Générale d'Editions, Paris.

Persky, J. (1972). 'Is the South a colony'? *Review of Radical Political Economics*, **1972**, (Winter).

Persky, J. (1973). 'The South: a colony at home', *Southern Exposure*, **1**, (Fall).

Pickvance, C.G. (1976). *Urban Sociology: Critical Essays*, Tavistock, London.

Pickvance, C.G. (1977). 'Marxist approaches to the study of urban politics: divergences among some recent French studies', *Int'l. Journal of Urban and Regional Research*, **1**, 219–255.

Pickvance, C.G. (1980). 'Policies as chameleons: an interpretation of regional policy and office policy in Britain', Urban and Regional Study Unit, University of Kent, Canterbury.

Pirenne, H. (1925). *Medieval Cities: Their Origins and the Revival of Trade*, Princeton University Press, Princeton, N.J.

Pirenne, H. (1958). *A History of Europe*, Vol. 1, Anchor Books, Garden City, New York.

Pirie, M. (1982). 'A short history of enterprise zones', *National Review*, **33/1**, 26–29.

Planque, B. (1977a). 'L'organisation spatio-economique de la France et de la R.F.A.', *La Diffusion Interrégionale du Développement*, Document de Recherche I.1, Centre d'Economie Régionale, Université d'Aix-Marseille III.

Planque, B. (1977b). 'Rapport de synthèse', *La Diffusion Interrégionale du Développement*, Centre d'Economie Régionale, Université d'Aix-Marseille III.

Popper, F. (1979). 'The capital eye', *Practicing Planner*, **9/2**, 11–13.

Poulantzas, N. (1976a). 'The capitalist state: a reply to Miliband and Laclan', *New Left Review*, **95**, 63–83.

Poulantzas, N. (Ed.) (1976b). *La Crise de L'Etat*, PUF, Paris.

Prats, Y. (1979). 'Vers une nouvelle approche de la planification régionale et urbaine en France?', *Cahiers d'Economie Politique*, **7**. Numéro special: 'Planification urbaine et régionale, vers de nouvelles approches' (February), pp. 51–57.

Pred, A. (1976). 'The interurban transmission of growth in advanced economies: empirical findings versus regional-planning assumptions', *Regional Studies*, **10**, 151–171.

Proceedings of the First National Conference on City Planning, Washington, D.C., 21–22 May 1909 (1910). USGPO, Washington, D.C.

Proudhon, P.-J. (1840) *Qu 'est-ce-que la Propriété? ou Recherche sur le Principe du Droit et du Gouvernment*, Paris. Current edition: 1966, Garnier-Flammarion, Paris.

Proudhon, P.-J. (1843). *De la Création de l'Ordre dans l'Humanité, ou Principes d'Organisation Politique*, Besançon. Current edition: 1959, *Oeuvres Completes*, Marcel Rivière, Paris.

Proudhon, P.-J. (1846). *Système des Contradictions Economiques ou Philosophie de la Misère*, 2 vols, Paris. Current edition: Collection 10–18, Paris.

Proudhon, P.-J. (1849). *La Banque du Peuple*, Paris. Current edition: 1959, *Oeuvres complètes*, Librairie Marcell Rivière, Paris.

Proudhon, P.-J. (1851). *L'idée Générale de la Révolution au XIXe Siècle*, Paris. Current edition: 1959, *Oeuvres Complètes*, Librairie Marcel Rivière, Paris.

Proudhon, P.-J. (1858). *De la Justice dans la Révolution et dans l'Eglise*, 3 Vols, Paris. Current edition: 1959, *Oeuvres Complètes*, Librairie Marcel Rivière, Paris.

Proudhon, P.-J. (1862). *La Fédération et l'Unité en Italie*, Paris. Current edition: 1959, *Oeuvres Complètes*, Librairie Marcel Rivières, Paris.

Proudhon, P.-J. (1863). *Du Principe Fédératif et de la Nécessité de Reconstituen le Parti de la Révolution*, Paris. Current edition: 1959, *Oeuvres Complètes*, Librairie Marcel Rivières, Paris.

Proudhon, P.-J. (1865a). *La Théorie de la Propriété*, Paris. Current edition: 1959, *Oeuvres Complètes*, Librarire Marcel Rivière, Paris.

Proudhon, P.-J. (1865b). *De la Capacité Politique des Classes Ouvrières*, Paris. Current edition: 1959, *Oeuvres Complètes*, Librairie Marcel Rivières, Paris.

Prud'homme, R. (1974). 'Regional economic planning in France', in *Public Policy and Regional Economic Development: The Experience of Nine Western Countries* (Ed. N.M. Hansen), Bellinger, Cambridge.

Purdom, C.B. (1925). 'New towns for old: garden cities—what they are and how they work', in L. Mumford (Ed.), 'The regional plan number', *Survey Graphic*, **54**, (May). Reprinted in C. Sussman (Ed.), 1976, *Planning the Fourth Migration*, MIT Press, Cambridge, Massachusetts.

Purdom, C.B. (1949). *The Building of Satellite Towns*, Dent, London.

Rafuse, R. (1977). 'The new regional debate: a national overview', National Governor's Conference, Washington, D.C.

Reclus, E. (1905–1908) *L'Homme et la Terre*, Librairie Universelle, Paris.

Rees, G., and Lambert, J. (1979). 'Urban development in a peripheral region: some issues from South Wales'. A paper presented to the Centre for Environmental Studies Conference on Urban Change and Conflict, University of Nottingham, 5–8 January 1979.

Rees, J., Hewings, G.J.D., and Stafford, H.A. (1981). *Industrial Location and Regional Systems*, J.F. Bergen, New York.

Regional Social Theory Workshop (1978). 'Accumulation, the regional problem and nationalism', in *London Papers in Regional Science* (Ed. P.W.J. Batey), **8**, 67–81.

Regional Studies (1982). Theme issue: 'The changing environment for regional policy in the United Kingdom', **16/5**.

Ricardo, D. (1953). *Principles of Political Economy and Taxation*, Cambridge University Press.

Richards, J., and Pratt, L. (1979). *Prairie Capitalism: Power and Influence in the New West*, McClelland and Stewart, Toronto.

Richardson, H.W. (1969). Regional Economics: Location Theory, *Urban Structure, and Regional Change*, Praeger, New York.

Richardson, H.W. (1972). 'Optimality in city size, systems of cities, and urban policy: a sceptic's view', *Urban Studies*, **9**, 29–48.

Richardson, H.W. (1973a). *Regional Growth Theory*, John Wiley, New York.

Richardson, H.W. (1973b). *The Economics of Urban Size*, Saxon House, Westmead.
Richardson, H.W. (1978a). *Regional and Urban Economics*, Penguin, Harmondsworth.
Richardson, H.W. (1978b). 'Growth centres, rural development and national urban policy—a defense', *Int'l. Reg. Sci. Rev., 3:2*, 133–152.
Richardson, H.W. (1979). 'Polarization reversal in developing countries', Paper presented to the North American Regional Science Association Conference, Los Angeles, November 1979.
Rifkin, J., and Barber, R. (1978). *The North Will Rise Again*, Beacon Press, Boston.
Robinson, J. (1979). *Aspects of Development and Underdevelopment*, Cambridge University Press, Cambridge, Massachusetts.
Robock, S.N. (1967). 'An unfinished task: a socio-economic evaluation of the TVA experiment', in *The Economic Impact of the TVA* (Ed. J.R. Moore), University of Tennessee Press, Knoxville.
Rodney, W. (1972). *How Europe Underdeveloped Africa*, Tanzania Publishing House, Dar-es-Salaam.
Rodwin, L. (1956). *The British New Towns Policy*, Harvard University Press, Cambridge, Massachusetts.
Rodwin, L. (1963). 'Choosing regions for development', in *Public Policy: Yearbook of the Harvard University Graduate School of Public Administration* (Eds C.J. Friedrich and S.E. Harris, Vol. 12, Harvard University Press, Cambridge, Massachusetts.
Rodwin, L. (1970). *Nations and Cities: A Comparison of Strategies for Urban Growth*, Houghton Mifflin, Boston.
Rodwin, L. (1978). 'Regional Planning in Less Developed Countries: A Retrospective View of the Literature and Experience', *International Regional Science Review, 3*, (2), pp. 113–131.
Rogers, A. (1979). *Economic Development in Retrospect*, V.H. Winston, Washington, D.C.
Roland, A. (1977). 'Impact de la politique des pouvoirs publics sur la localisation des industries en France', *La Diffusion Inter-régionale du Développement*, Document de Recherche II.1, Centre d'Economie Régionale, Université d'Aiz-Marseille III.
Roosevelt, F.D. (1938). *The Public Papers and Addresses of Franklin D. Roosevelt*, Vol. 1, Random House, New York.
Rosanvallon, P. (1976). L'Age de l'Autogestion, Editions du Seuil, Paris.
Ross, G.W., and Cohen, S. (1975). 'The politics of French regional planning', in *Regional Policy* (Eds J. Friedmann and W. Alonso), MIT Press, Cambridge.
Rostow, W.W. (1961). *The States of Economic Growth: A Non-Communist Manifesto*, Cambridge University Press.
Roterus, V., and Wesley, C. (1955). 'Notes on the basic-nonbasic employment ratio', *Econ. Geog., 31* (January), 17–20.
Rothblatt, D.R. (1971). *Regional Planning: The Appalachian Experience*, Heath Lexington Books, Lexington, Massachusetts.
Rowthorn, R., and Hymer, S. (1971). *International Big Business*, University Press, Cambridge.
Rudé, G. (1964). *Revolutionary Europe, 1783–1815*, Meridan Books, Cleveland.
Sant, M.C. (1975). *Industrial Movement and Regional Development: The British Case*, Pergamon Press, Oxford.
Sax, E. (1869). *The Housing Conditions of the Working Classes and Their Reform*, Vienna.
Schlesinger, A.M., Jr. (1965). *A Thousand Days: John F. Kennedy in the White House*, Fawcett Publications, Greenwich, Conn.
Seers, D. (1977). 'The new meaning of development', *International Development Review, 3*, 2–7.
Seers, D., Schaffer, B., and Kiljunen, J.L. (Eds) (1978). *Underdeveloped Europe: Studies in Core–Periphery Relations*, Harvest Press, Hassocks, Sussex.

Selznick, P. (1949). *TVA and the grass roots*, University of California Press, Berkeley and Los Angeles.

Shapley, D. (1976). 'TVA today: former reformers in an era of expensive electricity', *Science*, **194**, 814–818.

Shearer, D. (Ed.) (1979). *Public Policy and the New Progressives*, School of Architecture and Urban Planning, University of California, Los Angeles.

Sidor, J. (1982). 'State enterprise programs', The Council for State Community Affairs Agencies, Washington, D.C.

Siebert, H. (1969). *Regional Economic Growth: Theory and Policy*, International Textbook Co., Scranton.

Sinclair, W. (1979). 'Innovation of the mish-mash? Congress considers network of regional commissions', *L.A. Times*, 15 June.

Sion, J. (1908). *Les Paysans de la Normandie Orientale*, Armand Colin, Paris.

Smith, A. (1776). *The Wealth of Nations*. Current edition: 1970, Penguin Books, Harmondsworth.

Smith, A. (1970). *The Wealth of Nations* (Edited by A. Skinner), Penguin, Harmondsworth.

Smith, D.M. (1981). *Industrial Location: An Economic Geographical Analysis*, 2nd ed., John Wiley, New York.

Social Policy (1979). 'Special issue on organizing neighbourhoods', September/October.

Soja, E. (1980). 'The socio-spatial dialectic', *Annals of the Association of American Geographers*, **70**, 207–225.

Sorre, M. (1913). *Les Pyrénées Méditerranéennes*, Armand Colin, Paris.

Spaeth, D. (1981). *Ludwig Karl Hilberseimer: An Annotated Bibliography and Chronology*, Garland Publishing, New York.

'Special feature on the North South Question' (1979). *The Professional Geographer*, **31/1**, 34–74.

Stalley, M. (Ed.) (1972). *Patrick Geddes: Spokesman for Man and the Environment*, Rutgers University Press, New Brunswick, N.J.

Stein, C.S. (1957). *Towards New Towns in America* (With an introduction by Lewis Mumford), MIT Press, Cambridge, Massachusetts.

Stephens, J., and Holly, B. (1980). 'The changing patterns of industrial corporate control in the metropolitan United States', in *The American Metropolitan System* (Eds J. Brunn and J. Wheeler), Halsted Press, New York.

Sternleib, G., and Hughes, J.W. (Eds.) (1975). *Post-industrial America: Metropolitan Decline and Inter-regional Job Shifts*, Center for Urban Policy Research, Rutgers University, New Brunswick, N.J.

Sternleib, G., and Hughes, J.W. (1977). 'New regional and metropolitan realities of America', *Journal of the American Institute of Planners*, **43/3**, 227–241.

Sternleib, G., and Listokin, D. (Eds) (1981). *New Tools for Economic Development: The Enterprise Zone, Development Bank and RFC*, Center for Urban Policy Research, Rutgers University, Piscataway, N.J.

Sternsher, B. (1964). *Rexford Tugwell and the New Deal*. Rutgers University Press, New Brunswick, N.J.

Stöhr, W.B. and Tödting, F. (1977) 'Spatial equity—some anti-theses to current regional development doctrine' in *Papers of the Regional Science Association*, **38**, 33–53.

Stolper, W. (1947). 'The volume of foreign trade and the level of income', *Quart. J. of Econ.*, **61** (February), 285–310.

Stretton, H. (1978). *Urban Planning in Rich and Poor Countries*, Oxford University Press.

Sundquist, J. (1975). *Dispersing Population: What America Can Learn from Europe*, The Brookings Institute, Washington.

Sunkel, O. (1970). 'Desarrollo, subdesarrollo, dependencia, marginancion y desigualdades espaciales: hacia un enfoque totalizante', *Revista Latinoamericana de Estudios Urbanos y Regionales*, **1**, 13–51.

Sunkel, O. (1973). 'Transnational capitalism and national disintegration in Latin America', *Social and Economic Studies*, **22**, 132–176.

Sussman, C. (Ed.) (1976). *Planning the Fourth Migration: The Neglected Vision of the Regional Planning Association of America*, MIT Press, Cambridge, Massachusetts.

Sweezy, P.M. (1950). 'The transition from feudalism to capitalism', *Science and Society*, **14:2**, 134–157.

Sweezy, P.M., *et al.* (1954). *The Transition from Feudalism to Capitalism: A Symposium*, Science and Society, New York.

Taylor, S. (1981). 'The politics of enterprise zones', *Public Administration*, **59** (Winter), 421–439.

Thompson, W.R. (1965). *A Preface to Urban Economics*, Published for Resources for the Future, Johns Hopkins Press, Baltimore.

Thompson, W.R. (1968). 'Internal and external factors in the development of urban economies', in *Issues in Urban Economics* (Eds H.S. Perloff and L. Wingo Jr.), Published for Resources for the Future, Johns Hopkins Press, Baltimore.

Thorngren, B. (1970). 'How do contact systems affect regional development', *Environment and Planning*, **2**, 409–427.

Thornley, J. (1981). *Workers' Co-operatives: Jobs and Dreams*, Heinemann, London.

Tornqvist, G. (1970). *Contact Systems and Regional Development*, Lund Studies in Geography, Series B, Human Geography No. 35, C.W.K. Gleerup, Lund, Sweden.

Townsend, A.R. (1980). 'Unemployment geography and the new government's "regional" aid', *Area*, **12/1**, 9–18.

Tugwell, R.G. (1932). *Mr. Hoover's Economic Policy*, John Day, New York.

Tugwell, R.G. (1936). 'Housing activities and plans of the Resettlement Administration', in *Housing Officials Yearbook*, pp. 28–34, National Association of Housing, Washington, D.C.

Tugwell, R.G. (1937). 'The meaning of the greenbelt towns', *New Republic*, **90**, 42–43.

Tugwell, R.G., and Banfield, E. (1950). 'Grass roots democracy—myth or reality?', *Public Administration Revue*, **10**, 47–59.

Tym, R. *et al.* (1982). *Monitoring enterprise zones*, 3 vols, Urban Land Economists, March.

Ullman, E.L. (1958). 'Regional development and the geography of concentration', *Papers and Proceedings of the Regional Science Association*, **4**. Reprinted in J. Friedmann and W. Alonso (Eds), *Regional Development and Planning*, 1964, MIT Press, Cambridge.

UN Statistical Office (1977). *Yearbook of National Accounts and Statistics 1976*, Vol. 1, Individual Country Data, United Nations, New York.

US Bureau of the Census (1960). *Historical Statistics of the United States, Colonial Times to 1957*, USGPO, Washington, D.C.

US Department of Commerce, Bureau of Census (1978). *Statistical Abstract of the United States, 1978*. Washington, D.C.: USGPO, p. 261: and US Department of Commerce, Bureau of Census (1975). *Statistical Abstract of the United States, 1975*. Washington, D.C.: USGPO, p. 261.

US Department of Commerce, Economic Development Administration (1972). *Program Evaluation: The Economic Development Administration Growth Center Strategy*, US Department of Commerce, Washington.

US Department of Commerce, Economic Development Administration (1974a). *The EDA Experience in the Evolution of Policy: A Brief History, September 1965–June 1973*, USGPO, Washington.

US Department of Commerce and Office of Management and Budget (1974b). 'Report to the Congress on the proposal for an economic adjustment program', Washington, D.C.

186

US Federal Reserve Bank, Kansas City (1952). 'The employment multiplier in Wichita', *Monthly Review, Tenth Federal Reserve District,* **37:9**.

USGAO (1982). *Revitalizing Distressed Areas through Enterprise Zones: Many Uncertainties Exist*, Prepared by the Comptroller General, Report to Congress, GAO/CED-82-78, 15 July.

USHUD Office of Development and Research (1982). *State Legislation Concerning Enterprise Zones*, Washington, D.C., 4 April.

US National Planning Board (1934). *Final Report, 1933–34*, USGPO, Washington, D.C.

US National Resources Committee (1935). *The Regional Factors in National Planning*, USGPO, Washington, D.C.

US National Resources Committee (1936–1943). *Regional Planning*, 10 Vols, USGPO, Washington, D.C.

Vallaux, C. (1906). *La Basse Bretagne*, Armand Colin, Paris.

Vance, R.B. (1935). *Regional Reconstruction: A Way Out for the South*, Foreign Policy Association, New York.

Veblen, T. (1899). *The Theory of the Leisure Class: An Economic Study of Institutions*, Charles Scribner's Sons, New York.

Vernon, R. (1959). *The Changing Economic Function of the Central City*, Area Development Committee of CED: New York.

Vernon, R. (1966). 'International investment and international trade in the product cycle', *Quarterly Journal of Economics*, **80**, 190–207.

Vico, G. (1744). *La Scienza Nuova*. Translated from the 3rd Italian ed. by T.G. Bergin and M.H. Fische as *The New Science of Giambattista Vico*, 1970, Cornell University Press, Ithaca, N.Y.

Vidal de la Blache, P. (1903). 'Tableau de la géographie de la France', in *Histoire de France* (Ed. E. Lavisse), Vol. 1, Hachette, Paris.

Vidal de la Blache, P. (1917). *La France de l'Est: Lorraine–Alsace*, Armand Colin, Paris.

Vidal de la Blache, P. (1921). *Principes de Géographie Humaine* (Ed. by E. de Martonne) (posthumous), Armand Colin, Paris.

Vigier, P. (1977). 'Régions et régionalisme en France au XIXe siècle', in *Régions et Régionalisme en France, du XVIIIe Siècle à Nos Jours* (Eds C. Gras and G. Livet), PUF, Paris.

Viner, J. (1952). *International Trade and Economic Development*, Free Press, Glencoe, Illinois.

Vining, D.R., and Kontuly, T. (1978). 'Population dispersal from major metropolitan regions: an international comparison', *International Regional Science Review*, **3**, 49–73.

Vining, D.R., and Strauss, A. (1977). 'A demonstration that the current decentralization of population in the United States is a clean break with the past', *Environment and Planning, A*, 1977, 751–758.

Vining, R. (1946). 'Location of industry and regional patterns of business cycle behaviour', *Econometrica*, **14** (January), 37–68.

Voyene, B. (1973). *Le Fédéralisme de P.-J. Proudhon*, Press d'Europe, Paris and Nice.

Walker, R. (1978). 'Two sources of uneven development under advanced capitalism: spatial differentiation and capital mobility', *Review of Radical Political Economics*, **10**, 28–38.

Wallerstein, I. (1974a). 'Dependence in an interdependent world: the limited possibilities of transformation within the capitalist world economy', *African Studies Review*, **17**, 1–27.

Wallerstein, I. (1974b). 'The rise and future demise of the world capitalist system: concepts for comparative analysis', *Comparative Studies in Society and History*, **16**, 387–415.

Wallerstein, I, (1976a). *The Modern World System*, Academic Press, San Francisco.

Wallerstein, I. (1976b). 'A world-system perspective on the social sciences', *British Journal of Sociology*, **27**, 343–352.

Watkins, A.J. (1979). 'Good business climates and the second war between the states'. Paper presented to the annual meeting of the North American Regional Science Association, 9–11 Nov. 1979, Los Angeles.

Weaver, C. (Ed.) (1981a). 'Human settlements, bureaucracy and ideology: the Yugoslav alternative. A seminar with Ernest Weissmann', *Occasional Papers*, Centre for Human Settlements, University of British Columbia.

Weaver, C. (1981b). 'Development theory and the regional question: a critique of spatial planning and its detractors', in *Development From Above or Below* (Eds W.B. Stohr and D.R.F. Taylor), John Wiley, London.

Webber, M.J. (1972). *The Impact of Uncertainty on Location*, MIT Press, Cambridge, Massachusetts.

Webber, M.M. (1964). 'The urban place and the nonplace urban realm', in *Explorations into Urban Structure* (Eds M.M. Webber *et al.*), University of Pennsylvania Press, Philadelphia.

Weber, A.F. (1899). *The Growth of Cities in the Nineteenth Century: A Study in Statistics*, Published for Columbia University, Macmillan, New York. Reprint: 1963, Cornell University Press.

Weber, M. (1905). *The City*, Current edition: 1958, Free Press, Glencoe, Illinois.

Weber, A. (1928). *Theory of the Location of Industries* (Trans. by C.J. Friedrich from the 1st German ed., 1909), University Press, Chicago.

White House, Office of the Presidential Press Secretary (1982a). 'The administration's enterprise zone proposal fact sheet', Washington, D.C., 23 March.

White House, Office of the Presidential Press Secretary (1982b). 'Transmittal of the Enterprise Zone Tax Act of 1982', Washington, D.C., 23 March.

White, M., and White, L. (1962). *The Intellectual Versus the City*, Harvard University Press and MIT Press, Cambridge, Massachusetts.

White Paper on Industrial and Regional Development, Cmnd. 4942, HMSO, London, March 1972.

William-Olsson, W. (1941). *Stockholms framtida utvickling*, Stockholm.

Williamson, J.G. (1965). 'Regional inequality and the process of national development: a description of patterns', *Economic Development and Cultural Change*, **13**, 3–45.

Wilson, E. (1940). *To the Finland Station: A Study in the Writing and Acting of History*, Doubleday, New York.

Wilson, T. (1982). 'The macro-economic background of regional policy', *Regional Studies*, **16/5**, 383–389.

Woodcock, G. (1962). *Anarchism: A History of Libertarian Ideas and Movements*, pp. 106–144, Meridian Books, New York.

Woodcock, G. (1972). *Pierre-Joseph Proudhon: His Life and Work*, Schocken Books, New York.

Woodcock, G., and Arakumovic, I. (1971). *The Anarchist Prince, Peter Kropotkin*, Schocken Books, New York.

Woodruff, A.M. (Ed.) (1980). *The Farm and the City: Rivals or Allies?*, Prentice-Hall, Englewood Cliffs, N.J.

Woofter, T.J. (1934). 'Tennessee Valley Regional Plan', *Social Forces*, **12**, 329–338.

World Almanac and Book of Facts, 1979, Newspaper Enterprise Association, New York.

Wright, H. (1925). 'The road to good houses', in L. Mumford (Ed.), 'The regional plan number', *Survey Graphic*, **54** (May). Reprinted in C. Sussmann (Ed.), 1976, *Planning the Fourth Migration*, MIT Press. Cambridge, Massachusetts.

Wright, H. (1926). *Report of the New York State Commission on Housing and Regional Planning to Governor Alfred E. Smith*, J.B. Lyon, Albany, N.Y. Reprinted in C.

Sussman (Ed.), 1976, *Planning the Fourth Migration*, MIT Press, Cambridge, Massachusetts.

Yale University Directive Committee on Regional Planning (1947). *The Case for Regional Planning, With Special Reference to New England*, Yale University Press, New Haven. Connecticut.

Zelinsky, W. (1977). 'Coping with the migration turnaround: the theoretical challenge', *International Regional Science Review*, **2**, 175–178.

Author Index

Where an author is referred to in a footnote, the page reference number is followed by a small 'n' with the note number in parentheses.

The use of the Latin *bis* ('twice'), *ter* ('Three times'), or *quater* ('four times'), indicates an allusion to an author in two or more passages on the same page.

190

Built Environment (1981), 105
Burchell, R.W., and Listokin, D. (1980), 8
Burns, S. (1975), 150, 157 n(8)
Business Week (1980), 140
Butler S.M. (1981), 99, (1982), 99
Buttell, F.H., and Newby, H. (1980), 5

Cahiers d'Economie Politique (1976), 102
Cameron, G.C. (1970), 92, (1974), 100
Carney, J.R. (1980), 128, 130, 134
Carney, J.R., Hudson, R., Ive, G., and Lewis, J. (1975), 122, 139 n(6)
Carney, J., Hudson, R., and Lewis, J. (1980), 7, 133 *bis*, 139 n(6), 146
Carney, M., and Shearer, D. (1980), 153
Carrère, P., Catin, M., and Lamandé, J. (1978), 8
Carroll, P.N., and Noble, D.W. (1977), 6, 26
Carter, I. (1974), 122
Case, J., and Taylor, R.C.R. (1979), 153
Cash, W.J. (1941), 11
Castells, M. (1973), 138 n(4)
Castells, M. (1978), 7
Chadwick, E. (1842), 31
Charles-Brun, J. (1911), 10, 45
Chase, S. (1925), 63, (1933), 63, 69, 74 n(5), (1936), 74 n(5)
Cherry, G.E. (1974), 34
Childe, V.G. (1936), 74 n(4)
Christaller, W. (1933), 73
Clark, C. (1938), 86
Clark, G.L. (1980a), 130, 139 n(6), 141, (1980b), 8, 131, (1981), 131
Cole, J.P. (1981), 129
Comte, A. (1852), 56 n(12), (1969), 56 n(12), (1972), 56 n(12)
Conkin, P. (1959), 66
Conrat, M., and Conrat, R. (1977), 6
Coraggio, J.L. (1977), 113
Crisis Reader Editorial Collective (1978), 7
Cullingworth, J.B. (1980), 90 n(1), 93
Cultiax, D. (1975), 131
Cumberland, J.C. (1971), 24, 92

Daley, M.C. (1940), 90 n(3)
Damesick, P. (1982), 104
Damette, F. (1980), 108
Damette, F., and Poncet, E. (1980), 108
Daniels, P.W. (1969), 100, (1977), 100

Darwent, D.F. (1969), 84
DATAR (1972), 11 n(5)
Davies, T. (1974), 139 n(6)
Deane, P. (1978), 143, 148
Dear, M. (1980), 139 n(6)
Dear, M., and Scott, A. (1980), 136
De Forest, R.W., and Veiller, L. (1903), 31
Delors, J. (1978), 102
Demangeon, A. (1905), 47
de Martonne, E. (1902), 47
Derthick, M. (1974), 69, 92 *bis*, 95, 98 *ter*, 11 n(6)
De Souza, A.R., and Foust, J.B. (1979), 129
De Souza, A.R., and Porter, P.W. (1974), 112
Dickinson, R.E. (1947), 60, (1964), 60
Dobb, M. (1946), 17
Doherty, J.C. (1980), 75 n(10)
Domhaff, G.W. (1979), 138 n(6)
Dore, R. (1973), 157 n(12)
Dulong, R. (1973), 157 n(12), (1967a), 133, 139 n(6), (1976b), 133, 139 n(6), (1978), 133, 139 n(6)
Dunbar, G.S. (1978), 42, 43
Duncan, O., and others (1960), 82
Dunford, M. (1977), 110 n(3), 131, (1979), 131
Dunford, M.F., Geddes, M., and Perrons, D. (1980), 93, 105, 139 n(6), 156 n(1), 141
Dusenberry, J.S. (1950), 90 n(3)

Editorial Collective (1978), 146
El-Shakhs, S. (1965), 84, (1972), 84
Emmanuel, A. (1972), 116
Engels, F. (1872), 55 n(3), (1873), 55 n(3), (1880), 55 n(3), (1887), 32, (1892), 31
Ewen, L.A. (1978), 170

Fanon, F. (1961), 129
Farallones Institute, (1979), 157 n(9)
Firn, J. (1975), 122
Flory, T. (1966), 45 *bis*, 46 n(6)
Fourier, C. (1808, 1841), 32, (1882), 32, (1829, 1848), 32, 34 *ter*, (1971), (trans. Julia Franklin), 6
Frank, A.G. (1967), 113, 146, (1971), 115 *bis*
Freed, D. (1980), 157 n(9)(10)
Freeman, T.W. (1961), 46

192

Hume, E., and Shannon, D. (1979), 13
Hymer, S. (1968), 5, 113, 114, (1972a), 5, 113, 114, 156, (1972b), 117, 156
Hymer, S., and Resnick, S. (1971), 21 bis, 114, 116
Hymer, S., and Rowthorn, R. (1970), 11, 116

IMF (1978), 174
Innis, H.A. (1930), 90 n(3), 145
Isard, W. (1956), 79, 80
Isard, W., and Peck, M.J. (1954), 4, 90 n(3)

James, P.E. (1972), 46
James, P.E., and Jones, C.F. (1954), 60
Jensen, J. (1951), 73
Jensen, H.T. (1980), 139 n(6)
Jessop, B. (1977), 138 n(6)
Johnson, D.A. (1983), 58
Jumper, S.R., Bell, T.L., and Ralston, B.A. (1980), 129

Kashdan, S. (1981a), 99 bis, (1981b), 99
Kellogg, P.U. (1914), 32
Kitchen, P. (1975), 49, 55 n(3)
Klaassen, L.H. (1965), 82, 86, 87, 141, 145, (1967), 86, (1968), 86
Klaassen, L.H., and Drewe, P. (1973), 87
Klaassen, L.H., Torman, D.H., van Dongen, and Koyck, L.J. (1949), 90 n(3)
Kropotkin, P. (1899), 44, 45, 55 n(6), (1902), 44
Kuklinski, A. (1972), 84, 145, (1975), 111 n(6)
Kuklinski, A., and Petrella, R. (1972), 84, 146
Kuznets, S. (1941), 86

Laclan, E. (1975), 138 n(6)
Lafont, R. (1967), 112, (1968), 122, (1971), 126, 146
Lagarde, P. (1977), 10, 23, 55 n(7), 93 bis, 109 n(2)
Langlois, J. (1976), 42
Lasuén, J.R. (1969), 84, (1974), 90 n(4)
Law, C.M. (1980), 100, 102, 111 n(8) bis
Lebas, E. (1979), 138 n(6)
Lefebvre, H. (1968), 136, (1970), 136, (1972), 136
Le Monde, (1981), 12, 103 bis
Lenin, V.I. (1917), 21
Le Play, F. (1877–1879), 48, (no date), 48

Levainville, J. (1909), 47
Leven, C. (1954), 90 n(3)
Levin, P. (1976), 176
Levy, J. (1979), 139 n(6)
Lewis, C.A. (1979), 152
Lewis, W.A. (1955), 115
Lilienthal, D. (1944), 69
Lipietz, A. (1977), 32, 146, (1979a), 8 bis, 100, 123, 133 (1979b), 100
Lipton, M. (1977), 112
Lojkine, J. (1977), 139 n(6)
Lonsdale, R.E., and Seyler, H.L. (1979), 132
Lösch, A. (1954), 80
Lower, A.R.M. (1933), 90 n(3)
Lubove, R. (1969), 23

Maarek, G. (1979), 130
McCallum, J.D. (1979), 93, 104 ter, 111 n(9)
McCarthy, K.F., and Morrison, P.A. (1979), 75 n(10)
McCarty, H., and Lindberg, J. (1966), 157 n(11)
McCrone, G. (1969), 111 n(6)
McDonald, I., and Howick, C. (1981), 108
MacFadyen, D. (1970), 34
McGee, T.G. (1971), 112
Machlup, F. (1943), 90 n(3)
MacKaye, B. (1928), 51, 58, 60, (1931), 63
Mackaye, B., and Mumford, L. (1928), 58
Mackintosh, W.A. (1928), 90 n(3)
MacLennan, D., and Parr, J.D. (1979), 102
Mairet, P. (1957), 50
Malonado-Dennis, M. (1972), 13
Malizia, E. (1978), 8, 123
Mandel, E. (1975), 5, (1968), 130, (1976), 130, (1978), 147
Mandle, J. (1978), 122
Manners, G., Keeble, D., Rodgers, B., and Warren, K. (1980), 102, 104, 111 n(9)
Markusen, A.R. (1978a), 8, 9, (1978b), 9, 130, 138 n(3), 139 n(6), 141
Markusen, A.R., and Gastrup, J. (1978), 124, 139 n(6)
Markusen, A.R., and Schoenberger, E. (1979), 131
Marquand, J. (1979), 100
Marsh, G.P. (1864), 60
Marshall, A. (1890), 148

Martin, C.H., and Leone, R.A. (1977), 109 n(1), 110 n(4)

Marx, K. (1847), 55 n(3), (1857a), 17, (1857b), 17, (1867), 17, (1875), 55 n(3), (1909), 130, (1965), 130, (1967), 143, (1976), 130

Marx, K., and Engels, F. (1846), 17, (1847), 17, (1970), 130, (1978), 3

Marx, K., Engels, F., and Lenin, V.I. (1972), 55 n(3)

Massey, D. (1974), 131, (1976), 131, 147 bis, (1978a), 130, (1978b), 4, 130, (1978c), 131

Massey, D.B., and Meegan, R.A. (1978), 131, (1979), 131

Mather, B., and Mather, M. (1978), 178

Mayeur, J.M. (1977), 45

Meier, G.M. (1953), 90 n(3)

Mera, K. (1973), 84

Merrington, J. (1975), 16, 17, 18, 19, 135, 136, 143

Meyer, J.R. (1963), 145

Michalet, C.A. (1975a), 117, 118, (1975b), 117, (1976), 11

Michon-Savarit, C. (1975), 131

Miller, S.M. (1978), 139 n(6)

Mingione, E. (1977), 110 n(3)

Moore, R. (1979), 121

Morgan, K. (1979), 108, 139 n(6), (1980), 139 n(6)

Moriarty, B. (1980), 132

Morison, J.E., Commager, H.S., and Leuchtenburg, W.E. (1977), 25 ter, 26, 30 n(7)

Morrill, R.L. (1979), 124

Morris, E., and Hess, K. (1975), 13, 152, 153 bis

Mosely, M.J. (1974), 84

Mounts, R. (1982), 123

Moynihan, D.P. (1978), 99

Mumford, L. (1925a), 7, 51, 58 bis, (1925b), 51, 63, (1938), 11, 34, 48, 51, 57, 58, 60 bis, 61, 63 quater, 65 bis, 69, 74, 75 n(11)(14), 90 n(4), (1966), 2, 49, 58, (1982), 179

Mumford, L., and Osborn, F.J. (1971), 90 n(1), (1972), 75 n(11)

Mumy, G.E. (1978–1979), 135, 136

Myhra, D. (1974), 65

Myrdal, G. (1957), 82, (1969), 142

Nabudere, D. (1977), 115

Nairn, T. (1977), 9, 12, 112

Nathan, R.P. (1978), 157 n(13)

National Urban League, (1982), 99

Neutze, G.M. (1965), 84

Newman, M. (1972), 98

New Towns Committee, (1946), 77

North, D.C. (1955), 4, 15, 79, 145, (1961), 24 ter, 25, 30 n(6), (1966), 24, 28

O'Connor, J. (1973), 4, 134, 139 n(6)

Odum, H.W. (1934), 59; 62, 65, (1936), 59, (1939), 62

Odum, H.W., and Jocher, K. (1945), 59

Odum, H.W., and Moore, H.E. (1938), 11, 48, 59, 65, 69

OECD (1970), 109 n(3), (1974), 109 n(3), 110 n(3), (1976), 110 n(3) bis

Offe, C. (1975), 138 n(6)

Offe, C., and Ronge, V. (1975), 138 n(6)

Osborn, J.F. (1918), 77

Osborn, F.J., and Whittick, A. (1963), 34, (1969), 90 n(1)

Owen, R. (1813), 32, (1972), 6

Paelink, J. (1965), 79

Palloix, C. (1975a), 132, (1975b), 132, (1975c), 132, (1977), 132, 147 bis

Panikkar, K.M. (1959), 30 n(4), 114

Parti Socialiste Français (1980), 12

Passeron, H. (1979), 11, 103

Pedersen, P.O. (1974), 82

Perloff, H.S., Dunn, E.S., Jr., Lampard, E.E., and Muth, R.F. (1960), 24, 28 ter, 79, 82, 85 bis, 143, 156 n(1)(2)

Perrin, J.C. (1975), 79, 93, 102 bis

Perroux, F. (1950), 79, (1955), 79, (1969), 11

Persky, J. (1972), 126, (1973), 126

Pickvance, C.G. (1976), 136, (1977), 139 n(6), (1980), 100

Pirenne, H. (1925), 15, 16, 17, (1958), 15

Pirie, M. (1982), 108

Planque, B. (1977a), 94, (1977b), 102

Popper, F. (1979), 99

Poulantzas, N. (1967a), 139 n(6), (1976b), 7, 134

Prats, Y. (1979), 102

Pred, A. (1976), 4

Proceedings of the First National Conference on City Planning, Washington, DC (1910), 32

Proudhon, P.-J. (1843), 34 ter, (1846), 40, 43, (1849), 41, (1851), 41 bis, (1858), 40, (1862), 41, (1863), 41, (1865a), 41, (1865b), 42

Proud'homme, R. (1974), 100

194

Subject Index

Where an indexable item appears in a footnote, the page reference number is followed by a small 'n' with the note number in parentheses.